Inventions & Inventors

Volume 5
Instruments & Measurement

About this Book

Today we are surrounded by technology that makes our lives easier, safer, and more exciting, and that allows us to do things our ancestors would find hard to believe. We can fly to other countries, speak to people miles away, see pictures sent from other planets, perhaps one day even visit them. Modern technology was not invented "out-of-the-blue," however—it has roots deep in the past. The high-tech telecommunications industry, for example, developed from inventions such as the 200-year-old telegraph, and the first rockets flew many centuries ago. This set of books tells the story of inventions that shape our lives, tracing their history and explaining how they worked then and how they work now.

Along the way special boxes cover particular subjects in extra detail: **Scientific Principles** boxes explain the theories and processes behind aspects of technology, and **Key Components** boxes let you view particular inventions in detail, learning what each part is called and what it does. **Society and Inventions** boxes highlight the effects that inventions had on society, for both good and bad. **Facts & Figures** boxes give you fascinating facts about both yesterday's and today's technological marvels.

You can use this reference set in a variety of ways. Each book covers a specific area of technology and is divided into chapters, which can be read on their own as separate stories. So, you can read up on air and space technology as a whole, or just pick out the chapter on jets. If you are not sure where to find an invention, look it up in the index. The index covers all ten volumes, so it can help you trace inventions and inventors throughout the set. "See also" lists at the end of each chapter guide you to places elsewhere in the set where you can read about related inventions. There are also lots of pictures and specially drawn artworks, which explain how things work.

In the text words in SMALL CAPITALS are links to topics that are covered in detail elsewhere— try looking them up in the index if you are interested in finding out more. Words in **bold** are explained in the glossary, which appears at the end of every book. The timeline at the end of the book covers all ten volumes and provides a valuable overview of the march of progress, covering technology from the first tools of millennia ago right up to computers—and shows you where to find them in the set. The Further Reading list helps you do your own research on inventions and technology.

Published 2000 by
Grolier Educational, Sherman Turnpike, Danbury, Connecticut 06816

© 2000 Brown Partworks Ltd

Library of Congress Cataloging-in-Publication Data

Inventions and inventors
        p.   c.m.
  Contents: v. 1. Air and space -- v. 2. Buildings, homes, and structures -- v. 3. Communications -- v. 4. Farming, food, and biotechnology -- v. 5. Instruments and measurement -- v. 6. Land and water transportation -- v. 7. Manufacturing and industry -- v. 8. Medicine and Health -- v. 9. Military and Security -- v. 10. Power and energy

Set ISBN 0-7172-9384-X (set : alk. paper)
Volume ISBN: 0-7172-9389-0

1. Inventions--History Juvenile literature. 2. Inventors--History Juvenile literature [1. Inventions--History. 2. Inventors.]
T15.I58 1999
609—dc21
                                    99–19310
                                        CIP

FOR BROWN PARTWORKS LTD
Consultant: Dr. Donald Franceschetti
Contributors: John Bassett, Richard Beatty, and Trevor Day
Project editor: Bridget Giles
Editors: Amanda Harman, James Kinchen,
  Chris King, and Anne O'Daly
Project development: Richard Beatty
Picture research: Susannah Jayes
Index: Kay Ollerenshaw

Printed in Singapore

Contents

Volume 5

Instruments & Measurement

Weights and Measures 4

Measuring Time...12

Scientific Investigation 22

Technology and Measurement 44

Surveying, Navigation, and Cartography............54

Timeline...64

Glossary 66

Further Reading..68

Set Index 69

75,442

Weights and Measures

The development of measurement and counting from ells to meters

◀ *An African weight, once used by the Asante people of southern Ghana and now displayed in the British Museum in London. Made of brass, it was used to weigh gold.*

The gradual evolution of human society brought with it a need for accurate, standardized measurement and counting systems. Early humans needed to be able to measure lengths consistently when making HUNTING TOOLS and building HOMES. As trade developed, these systems of measuring goods increasingly needed to be standardized to avoid disputes. Early forms of measurement usually related to the human body—in particular the arm. Many of these measurements, such as the foot and the inch, are still widely used today. Others, such as the ell, have fallen into disuse.

Imperial systems

Measurements using the human body often led to disputes because people were different shapes and sizes. One of the earliest recorded attempts to standardize the units came from Huang-Ti, a Chinese emperor who ruled around 2600 B.C. His aim was that any units used would be based on the length of his body and no one else's. Other monarchs also developed a system along similar lines. For example, Henry I of Saxony, in Germany, had his scepter made to be the length of his own imperial standard ell in A.D. 900. So the units of measurement changed with each

Using the body for measurements

The length of the thumb = **1 in (2.54 cm)**

The length of the foot = **1 ft (30.48 cm)**

The distance between the tip of the nose and the tip of the middle finger = **1 yard (91.44 cm)**

The height of a man or the width of the arms extended out to the sides
= **1 fathom (6 ft, or 1.83 m)**

The breadth of a finger
= **1 digit (¾ in, or 1.9 cm)**

The distance across the base of four fingers
= **1 palm (3 in, or 7.6 cm)**

The distance between the tip of the middle finger and the elbow = **1 cubit (18 in, or 45.7 cm)**

The distance between the elbows when the arms are held down by the sides of the body
= **1 English ell (45 in, or 1.14 m)**

The distance traveled in one step = **1 step**

The distance traveled in two steps = **1 pace**

Racing distances

In **ancient Greece** athletic running tracks were housed in buildings called stadiums. The ancient Greeks and **Romans** used the length of the first foot-race course (which was 600 Greek feet, or 625 Roman feet long) in the stadium at Olympia, southern Greece, as a measurement in NAVIGATION and ASTRONOMY.

When King Iphitus revived the Olympic games in 884 B.C. it was found that the new stadium was shorter than the original one in Olympia. According to the Greek writer Plutarch, this was because the original Olympian stadium was measured out by the mythological Greek hero

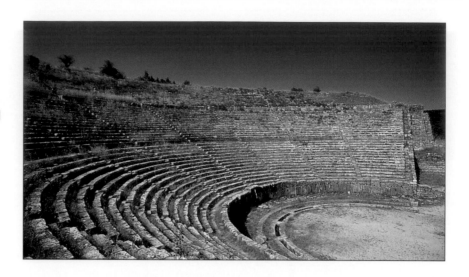

▲ *The Dodona amphitheater in Greece, a setting for ancient athletic events.*

Hercules, whose stride was longer than that of any other human being.

▶ *A painting of Frankish king and emperor Charlemagne (742–814), who made the size of his foot a standard imperial measurement around* A.D. *800.*

ruler. The confusion was greatest in Germany, where each county had its own set of measurements based on the bodily proportions of the various dukes and nobles.

Measuring weights

As trading increased, it was important to ensure that people were being sold the correct amount of goods. The Egyptians used their hands to measure out GRAIN: for example, one unit of volume was equal to two handfuls of grain. They also used cups and buckets as measures. Dutch colonists used the carat as a unit of weight for GOLD, with one carat equal to the weight of a single seed from the locust tree. The basic unit of weight in Britain was the grain, which was based on the weight of a grain of barley. In this system 7,000 grains were equal to one avoirdupois (British) pound. However, some trades had their own weight systems. For example, the troy pound, which was used by chemists, weighed 5,760 grains, while the pound used for weighing wool weighed 6,992 grains. The names of weights were all derived from Roman names, such as *unicia* for *ounce*.

Counting systems

▼ *The carob seeds of the locust tree, shown here, were once used as a measure of gold and precious stones. The unit of measure called the carat, which we still use today, comes from the Arabic word* quirrat, *meaning seed. The modern carat is equal to 0.00705 oz (0.2 g).*

Number systems have developed in every language, although the systems used often differ. The first written numbers date to 3000 B.C. and were discovered in Susa and Uruk in Iraq. Babylonian number systems developed between around 2200 and 1350 B.C. and were based on the number 60. The Mayan system was based on the number 20: the number of our fingers and toes.

The system of using 20 rather than 10 as a base existed in many cultures and languages. For example, it is possible to see that the French used this system by the fact that the French for 80 is *quatre-vingts* (which means *4 x 20*).

In the 5th century B.C. the Greeks used letters to represent their numbers. The first nine letters of the alphabet were used, with a space for a zero. The Romans also used letters to represent numbers:

- I = 1
- V = 5
- X = 10
- L = 50
- C = 100
- D = 500
- M = 1,000

Other numbers were denoted by simple combinations of these numerals. For example, the number six was denoted by VI (one more than five), while nine was IX (one less than ten). Unlike most other ancient

Scientific Principles

Binary code

The number of different single digits used in a number system is called its base. For example, in the decimal system the base is 10, and in the octal system the base is 8. In the system called **binary code** the base is 2 because there are only two different digits used: 0 and 1. Combinations of these binary digits (called bits for short) are grouped into bytes, which represent characters such as letters (**1**) and numbers (**2**) and are used as data and instructions in COMPUTERS.

In binary code the digits of a byte represent (from right to left) ones, twos, fours, eights, sixteens, and so on. The binary code for 106 is : 01101010 (64+32+8+2). (In the decimal system digits of a long number represent ones, tens, hundreds, thousands, and so on.)

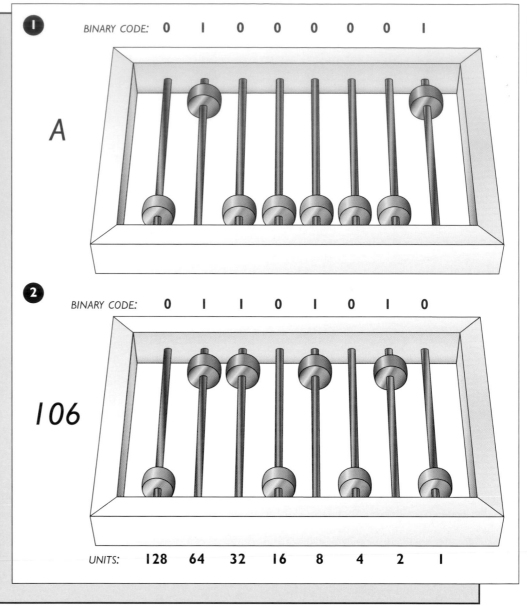

ways of denoting numbers, Roman numerals are still in widespread use in the Western world today. For example, we refer to Super Bowl XXVI rather than Super Bowl 26.

The Romans were also responsible for the introduction of the plus sign, which was a convenient, shorthand way of writing the Latin word *et* (meaning *and*).

The first recorded use of a **decimal** system of counting appeared in India in the 5th century A.D. In 829 an Arab mathematician from Baghdad called Muhammad ibn Musa al Khwarizmi (c. 800–c. 850) published

a book on **algebra** in which he used the decimal system. A French monk called Gerbert was responsible for the spread of Arabic number systems, which he discovered on a voyage to Cordoba in Spain in A.D. 980. He became Pope Sylvester II in 999 and used the new numbering system within the church. In 1202 Italian mathematician Leonardo Fibonacci (c. 1170–c. 1250) published his book *Liber Abaci*, which contained theories on numbers and mathematics based on the Arabic system. This led to the adoption of the Arabic system in many countries in Europe.

▶ *Pope Sylvester II, who as a young monk helped to spread Arabic numbering systems throughout the Western world.*

▼ *Like the Mayan people, the ancient Incas in Peru used a system of numbering based on knotted strings. The threads pictured here are now displayed in the Archeological Museum at the Peruvian capital of Lima.*

The ancient Babylonians used a circular symbol in their numbers, but zero—as a number in its own right—did not appear until the fourth century B.C. in India. The word *zero* comes from the Sanskrit word *sanya*, which means *nothing*.

Calculating and recording an amount of goods was originally done by means of a tally system in which a number of items would represent the goods sold. The Mayan civilization used a complex system of knotted threads to represent the numbers of people involved in the construction of bridges and buildings.

The abacus dates from 3000 B.C. and was first used in ancient Babylon. The name *abacus* comes from the Jewish word for *dust* because the original device was a slab of wood covered in sand on which the figures were written with an instrument called a stylus. However, the abacus eventually developed into a series of beads on a wire frame, with each column of beads representing a different set of numbers. It is possible to perform all of the four basic mathematical operations—addition, subtraction, division, and multiplication—on an abacus.

The birth of the metric system

British inventor James Watt (1736–1819) found the imperial system of measurement complex, and in 1783 he proposed a simple method of measuring using the decimal system, which is based on the number 10. Watt's system included the measurements meter, liter, and gram for length, volume, and weight.

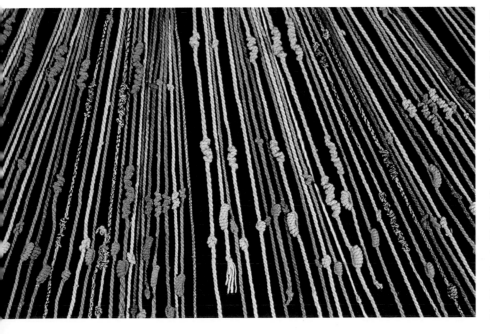

In 1790 French statesman Charles Maurice de Talleyrand (1754–1838) called on the French Academy of Sciences to develop a new system of measurement that would be uniform the world over. The proposed standard measurement of length was to be one ten-millionth of the distance between the north pole and the equator, on a line passing through Paris. This new unit was called the meter. All fractions or multiples of the new unit were to be based on the decimal system and counted in tens. On June 17, 1792, two French astronomers called Delambre and Mechain began measuring an area from Dunkirk to Barcelona on the MERIDIAN LINE that runs through the Paris observatory. Seven years later their results were presented to the

▶ *An engraving from 1795 showing the uses of the metric system: measuring the volume of liquids (1); weighing solid goods (2); and measuring lengths (3 and 4). The introduction of the metric system went hand in hand with the decimalization of money (5).*

Facts & Figures

● Although a grain of barley was once a common weight for measuring out goods, money was valued in relation to the weight of wheat. Four grains of wheat were equal to three grains of barley.

● The first scales and balances to measure weights were used in Egypt around 2500 B.C. and were used exclusively for weighing gold. Other trades adopted weighing scales some 1,200 years later.

● The kilogram (kg) is the unit of weight used in the metric system and was originally defined as the mass of one liter of water at 4°C.

● In a **vacuum** the speed of light is constant, at 186,281 miles (299,792,458 m) per second. This has given rise to a measurement in astronomy called the light-year, which is the distance traveled by light in a vacuum in one year (5.88 trillion miles, or 9.46 trillion km).

Key Components

The micrometer gauge

A micrometer gauge is a hand-held instrument used by scientists and engineers to take very precise measurements of objects. The key parts of a micrometer are the anvil and spindle—which hold in position the object being measured—and the barrel, which is marked with a Vernier scale. A ratchet is turned to tighten the spindle so that it grips the object as tightly as possible. The Vernier scale was invented by French mathematician Pierre Vernier (c. 1580–1637) and consists of a finely graded scale that is slid along a main scale to give tiny fractions of a measurement. Sometimes astronomers use a type of micrometer attached to their TELESCOPES to measure the distance between stars.

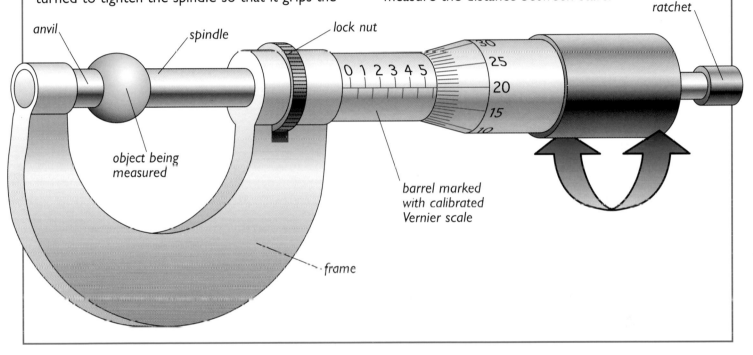

anvil

spindle

lock nut

ratchet

object being measured

barrel marked with calibrated Vernier scale

frame

French National Assembly, and on June 22, 1799, the first meter was presented to the people of France. In order to prevent any future dispute about the true length of a meter, the original platinum rod that defined the length was buried beneath the rock of a castle at Breteuil.

In 1960 the 11th General Conference on Weights and Measures met to establish a new definitive system of measurements that would be recognized the world over: the International System of Units. The conference gave new definitions for seven basic units for international use (known as SI units) on which all other units would be based. These were: the meter (m, used as a unit of length); kilogram (kg, weight); second (s, time); Kelvin (K, **thermodynamic** temperature); ampere (A, ELECTRIC CURRENT); mole (mol, amount of substance); and candela (cd, luminous intensity).

See also:
➤ The Age of Electricity (Vol. 10)
➤ Capturing Sound (Vol. 3)
➤ Computers (Vol. 3) ➤ Measuring Time (Vol. 5) ➤ Technology and Measurement (Vol. 5)

Measuring Time

The development of accurate time measurement, from the phases of the Moon to atomic clocks

◀ An hourglass, a simple means of measuring short periods of time. When the glass is turned upside down, grains of sand slowly begin to trickle from one chamber to the other through a small hole. The top chamber always takes the same amount of time to empty.

Today we are able to measure time very precisely, right down to the nearest fraction of a second. In early human society, however, measurements of time were concerned with much longer periods—days, months, and years. These measurements were based on natural cycles that repeated themselves at regular intervals. The 24-hour day, for example, is based on the rotation of the Earth. A point on the Earth turns to face the Sun and then turns away once in every 24 hours, giving rise to day and night. A lunar month (29.5 days) is based on the time for one complete lunar cycle, from one full moon to the next. The year is based on changes in day length, which in turn depend on the movement of the Earth about the Sun. These physical changes are mirrored by changes in the living world. Insect-eating bats leave their roosts at night and return before daylight, while tulip flowers open during the day and close at night. In spring leaves grow afresh; in the fall they drop to the ground.

Days and years

Many early peoples were fascinated by the regular movement of the stars, Sun, and Moon. They plotted their movements year after year, and certain alignments of celestial objects and key turning points in the year were marked by special celebrations. Temples were built to mark and predict these regular changes.

▶ Migrating birds, such as snow geese, are highly attuned to the changing seasons. Instinct tells them exactly when to begin their annual journey to warmer climates.

◀ *Sunrise on Midsummer Day at Stonehenge. Some people believe that this prehistoric monument may have acted as a primitive calendar based on the movements of the Sun and Moon.*

STONEHENGE, a circular arrangement of stones on Salisbury Plain in England, was used in ancient times as an observatory and as a focus for ceremonies. Construction began around 2300 B.C. and was finished c. 1400 B.C. An ancient avenue leading to Stonehenge is aligned with the position of the rising Sun at the summer solstice (the day of the year with the most hours of daylight).

Perhaps the most obvious regular change in the sky is the movement of the Sun in a curve, rising in the east and setting in the west. The height of the curve changes with the seasons: in summer it is higher in the sky, and so there are more hours of daylight; in winter the curve is lower, and the hours of daylight are fewer. The seasonal changes are caused by the rotation of the Earth around the Sun.

The ancient Babylonians, who lived between about 2150 and 500 B.C., were very knowledgeable about the seasons. During this time they built step pyramids called ziggurats

from which they observed the Moon and the stars. They were among the first true astronomers.

The Babylonians counted with a system based on the number 12, and it is they who probably divided day and night into 12 divisions each, so producing the 24 hours that make up

▼ *The California grunion fish spawns on beaches at full moon in March when the tides are at their highest.*

Scientific Principles

The seasons

The Earth moves in an **orbit** around the Sun, making one full revolution in a year. The seasons arise because the axis of the Earth (the line about which it rotates) is always tilted at just over 23°. In summer the Northern Hemisphere is tilted toward the Sun, and so the Sun is higher in the sky and hours of daylight are longer. In winter the Northern Hemisphere is tilted away from the Sun, so hours of daylight are shorter.

The longest day of the year in the Northern Hemisphere is midsummer (around June 21). This is also known as the summer solstice. The shortest day (midwinter or the winter solstice) is on December 21. Twice a year, on March 21 and September 23, the length of the day is equal to that of the night (12 hours). These days are known as the vernal (or spring) equinox and the autumnal (or fall) equinox respectively.

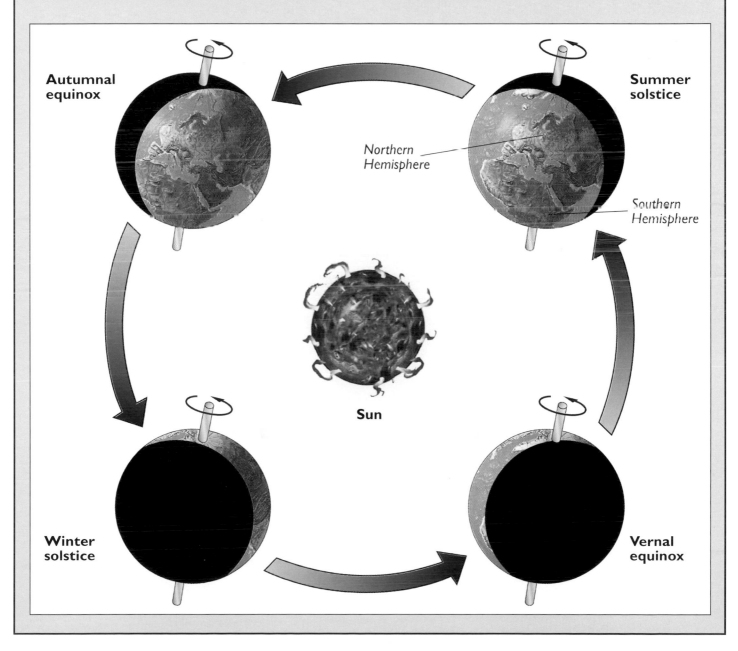

a complete day. However, in those ancient times an hour was simply one-twelfth of the amount of daylight or darkness. Its length varied with the seasons. For example, a "light-hour" was longest in midsummer and shortest in midwinter.

Weeks and months

Ancient peoples also noticed that the Moon showed a regular cycle, called the lunar cycle. The lunar cycle plays an important role in the lives of animals and plants. The **gravitational** effect of the Moon means that tides are highest at new moon and full moon. For many marine creatures important cycles are timed to coincide with full moon and the strongest tides. For example, off western Australia mass spawning of coral occurs in March, a few days after the full moon and highest tides. In California, again in March around the full moon and highest tides, the grunion fish spawns on sandy beaches. Two weeks later, when the highest tides return, the eggs hatch.

There is no straightforward relationship between lunar cycles and the seasons. So when ancient peoples were designing calendars, they found that they could not make lunar months fit neatly with the solar year based on the seasons.

The calendar

By 1000 B.C. at least two calendars were well established. The Babylonians used a calendar based on a lunar year (12 lunar months). Since each lunar month is 29.5 days, they made their months alternate between 29 days and 30 days. Twelve lunar months adds up to

Waxing or waning?

During the first half of the lunar cycle the Moon is said to be waxing, or growing bigger; during the second half of the cycle it is waning, or getting smaller. There is an easy way to tell whether the Moon is waxing or waning. If the Moon's crescent is to the right, like the curve in a capital letter "D," then the Moon is waxing. If the crescent is to the left, as in a capital "C," then the Moon is waning.

354 days, so to make the lunar years keep in step with the solar year (around 365 days), they had to add in an extra lunar month every few years. So some years had 12 lunar months and others had 13.

The Egyptians used a calendar based on a solar year. They noticed that the brightest star in the sky—the Dog Star, or Sirius—reappeared low in the eastern sky just before sunrise, and within days of its appearance the Nile River would flood. This combination of events was used to

◀ *Pope Gregory XIII, from whom the Gregorian calendar gets its name. It was Gregory who corrected the flaws in the earlier Julian system and created the calendar that the Western world uses to this day.*

Scientific Principles

Phases of the Moon

At the start of a lunar cycle the Moon is invisible and is known as a new moon (**1**). The Moon is next seen as a brightly lit crescent (**2**) that grows over seven days to become a half-moon (**3**). This stage is the first quarter of the lunar cycle. Over the next seven days the brightly lit part of the Moon continues to grow in size until the entire Moon is lit (**3** to **5**). This stage is called the full moon, and it happens after about 14 days, or halfway through the cycle. Over the next seven days the illuminated part of the Moon gets smaller until only half the Moon is lit—this is called the last quarter. Finally, in the last seven days of the cycle the remaining crescent gradually shrinks to nothing (**5** to **1**). The entire cycle lasts 29.5 days, which is the length of a lunar month (four lunar weeks). This means that each quarter of the lunar cycle lasts about one week.

The phases of the Moon happen because the Moon orbits around the Earth once every 29.5 days. When the Moon is between the Sun and the Earth, we see a new moon because the side of the Moon facing the Earth is in shadow. At full moon the Sun and Moon are on opposite sides of the Earth, and the side facing the Earth is fully lit by the Sun.

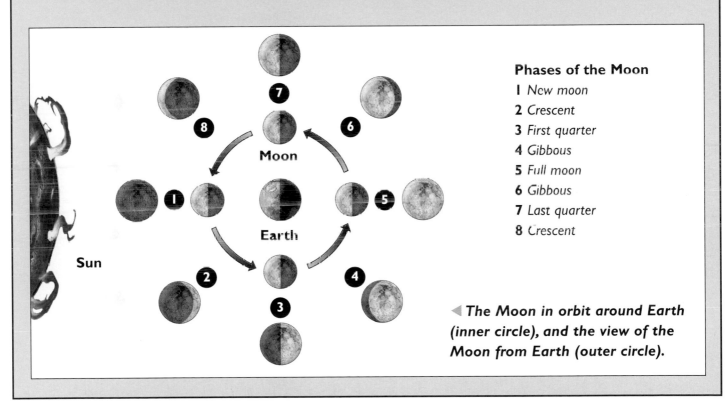

Phases of the Moon
1 *New moon*
2 *Crescent*
3 *First quarter*
4 *Gibbous*
5 *Full moon*
6 *Gibbous*
7 *Last quarter*
8 *Crescent*

◄ *The Moon in orbit around Earth (inner circle), and the view of the Moon from Earth (outer circle).*

fix the calendar, and so the Egyptians arrived at a year of 365 days.

By the **Middle Ages** many kinds of calendar were being used. Several of them—including the Hebrew, Islamic, and Chinese calendars—remain to this day. Most countries in the Western world, however, use a calendar based originally on that introduced in 46 B.C. during the rule of Roman Emperor Julius Caesar (c. 101–44 B.C.). The so-called Julian calendar remained in widespread use for over 1,500 years and gave rise to the months of the year we recognize today. All months except February have 30 or 31 days. February has 28 days, except once every four years (in a leap year), when it has 29. This calendar worked well. However, a solar year is actually 365 days, 5 hours, 48 minutes, and 46 seconds, and the Julian calendar is based on a year being around 11 minutes and 14 seconds longer than this. Over many centuries the Julian calendar drifted several days from the natural calendar based on the movement of the Sun and the seasons of the year. Around A.D. 1580 Pope Gregory XIII (1502–1585) was told by his astronomers how to correct this problem. So, in 1582 he dropped ten days from October, so that October 4 was followed by October 15. Most other Roman Catholic European countries followed his example. Britain and the U.S. colonies, however, did not change to the so-called Gregorian calendar until 1752. The Gregorian calendar has a leap-year day in century years that are divisible by 400—such as 1200, 1600, and 2000. With the Gregorian calendar the difference between the calendar year and the solar year is only around 26 seconds. It is the Gregorian calendar that Western countries use today.

The earliest clocks

The movement of the Sun was used as the basis for some of the earliest clocks. Sundials and shadow clocks measure time using the shadow of a marker (called the gnomon) cast by the movement of the Sun. The shadow falls onto a scale that measures elapsed time. Sundials have their limitations, however. The position of the shadow varies with the season of the year, while the clock cannot be used on a cloudy day or at night. The earliest shadow clocks date

▲ A vertical sundial mounted on a wall. As the Sun moves across the sky during the day, the shadow moves around a scale etched into the base of the sundial and tells the time.

from about 1500 B.C. and were used in ancient Egypt. Sundials were also common in **ancient Greece and Rome**. The Romans even used portable sundials—the ancient equivalent of the modern wristwatch. Sundials stayed in common use as late as the 19th century, until cheap and accurate mechanical clocks replaced them.

Flowing or dripping water was also used as a means of measuring time. The earliest water clocks date from about 1400 B.C. in ancient Egypt. They look rather like flowerpots with tiny holes in the bottom, through which water escapes. Time is measured as the pot gradually empties. Much more sophisticated water clocks—with LEVERS and PULLEYS

Facts & Figures

● There are several absolute measures of time. Time based on the movement of the Sun and the seasons is called universal time (UTO). Time measured by the cesium atomic clock is atomic time (AT). Time based on both universal and atomic time is called universal coordinated time (UCT).

● Noon (when the Sun is at its highest in the sky) in New York does not occur at the same time as noon in London, which is around 5,000 miles (8,000 km) to the east. One revolution of the Earth happens every 24 hours, so 15° of turn is equivalent to one hour (360 divided by 24). London is 74° east from New York, which means that New York time is about 5 hours behind London. To deal with this problem, the world is divided into time zones, all measured relative to the time at Greenwich in London (called Greenwich Mean Time, or GMT).

● Today, tall pendulum clocks are commonly known as grandfather clocks. They got their name from a song called *My Grandfather's Clock*, which was popular in the late 19th century. Before this they were often referred to as coffin clocks since they were often concealed in a 6-ft (2-m) wooden box.

powered by rising water—were invented by the Greeks and Romans.

The sandglass, or hourglass—which uses sand trickling through a narrow opening—was used in the Middle Ages for measuring short periods of time. Burning candles and smoldering lengths of knotted rope were also used to measure time intervals.

Mechanical clocks and watches

In 1581 Italian astronomer and physicist Galileo Galilei (1564–1642) noted that a freely swinging pendulum marks out a regular time that depends on the length of the string. A shorter string marks out a shorter time period than a longer string. Using this principle, the first pendulum clocks were made in the mid-17th century by Dutch physicist and astronomer Christiaan Huygens (1629–1695). Pendulum clocks were the most popular—and expensive— clocks in Europe until the mid-18th

▶ *Christiaan Huygens presents one of his pendulum clocks to his patron, King Louis XIV of France, in 1659. Huygens designed the first pendulum clock on Christmas Day in 1656.*

Key Components

Pendulum clocks

Pendulum clocks work because the amount of time that a pendulum takes to swing from side to side is constant and depends on its length. A pendulum 39 in (99 cm) in length will swing from side to side exactly once every second.

In a pendulum clock the pendulum is linked to a mechanism called a pallet. When the pendulum swings to the right, a tooth on the left-hand side of the pallet catches on a toothed wheel called the escape wheel. The escape wheel is rotated by a weight attached to a spindle. The pallet temporarily stops the rotation of the wheel.

When the pendulum starts to swing to the left, the wheel is released, only to be stopped by the right-hand tooth of the pallet a second later.

A series of GEARS and spindles links the escape wheel to hands on the clock face. In this way the hands are allowed to move clockwise once every second, so recording the passing of time.

▶ *The key components of most modern wristwatches are quartz crystals, which use the PIEZOELECTRIC effect to power the watch.*

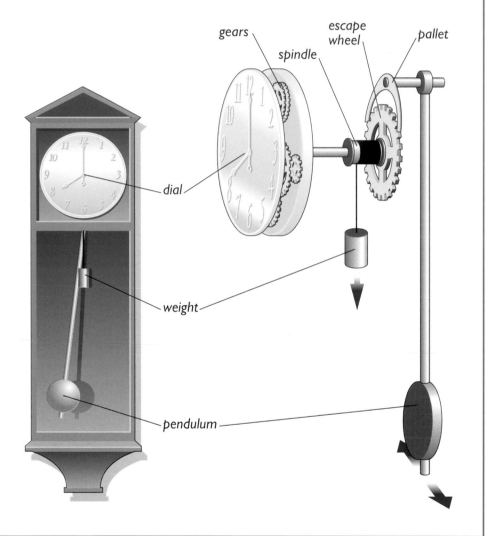

gears

spindle

escape wheel

pallet

dial

weight

pendulum

century, when spring-driven clocks were introduced. Among the first highly accurate clock of this type was the ship's clock (called a CHRONOMETER) designed by British watch- and clockmaker John Harrison (1693–1776) in 1761. It was accurate to within a few seconds a week.

Electronic and atomic clocks

As early as the 1920s clockmakers realized that crystals of a mineral called quartz could be harnessed to measure intervals of time. Quartz crystals are **piezoelectric**, which means that they vibrate at a particular frequency when an ELECTRIC CURRENT

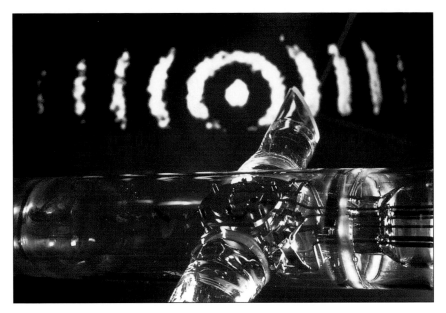

developed the first commercial quartz watches in 1969.

In 1948 a team of scientists, including U.S. chemist Willard F. Libby (1908–1980), worked together on a project to use the vibration of cesium atoms to mark out time with great accuracy. The wave energy produced by cesium atoms vibrates at nearly 9.2 billion cycles per second. In fact, the second is now officially defined in this manner. Modern cesium atomic clocks have an accuracy of one second in ten thousand billion—they lose around one second in 300,000 years.

▲ *A laser beam entering part of an atomic clock. This clock uses cesium atoms and is accurate to one billionth of a second per day.*

excites them. The vibration is detected and is counted off on a watch or clock face. The first prototype quartz clock was constructed by U.S. electrical engineer Warren Alvin Marrison (1896–1980) in 1929. The Seiko watch company

See also:

➤The Age of Electricity (Vol. 10)
➤Surveying, Navigation, and Cartography (Vol. 5) ➤Technology and Measurement (Vol. 5) ➤Weights and Measures (Vol. 5)

Society and Inventions

Einstein's theory of relativity

Perhaps the world's most famous physicist, Albert Einstein's (1879–1955) greatest achievement was his theory of relativity, which he conceived in 1905. One of its conclusions was that time is relative. At very high speeds—those nearing the speed of light (186,300 miles or 299,792 km per second)—time would seem to slow down relative to those moving at much slower speeds. This being the case, astronauts traveling across space at high speeds would age more slowly than the friends and relatives they had left behind on Earth.

Scientific Investigation

How the need for accurate measurements in science has driven the development of many instruments of ever-increasing sophistication

◀ Telescopes were originally simple sets of aligned glass lenses but have now evolved into instruments of incredible power and complexity. From its orbit around the Earth the Hubble Space Telescope (shown here) can collect light from distant galaxies and stars without distortion from the atmosphere.

Scientists have developed a huge range of measuring instruments to learn more about the natural world. Some are passive—they examine phenomena that happen naturally—while others are active—they cause some kind of experimental change in the system being studied in order to assess its effects. Many of the greatest pioneering scientists had to invent their own instruments.

Optical instruments

The easiest way to find out about anything is simply by looking at it. The eye itself is a passive instrument designed to gather light from our surroundings and focus it onto a detector (nerve cells at the back of the eye). However, often it is not possible to gain much information by just looking at an object because it is either too small or too far away. Even those objects that can be seen with the eye alone may need to be looked at in more detail. To do this, scientists use two main types of instrument—microscopes and telescopes.

Microscopes are used to look at objects close up, while telescopes are designed for looking at distant objects.

▶ A selection of Robert Hooke's inventions for scientific investigation, including his microscope (bottom).

The two instruments work by the same principle: they appear to bring an object much closer to the observer.

Microscopes

A microscope uses lenses to collect light rays as they spread out from different points on an object and bring them together in an image, either on a screen or in the eyes of the observer. The simplest magnifiers,

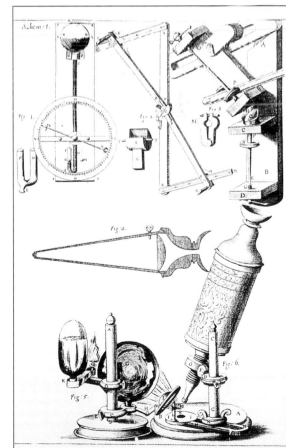

invented by eyeglass-makers in Holland around the 1590s, employed only a single lens and were all that Dutch biologist and engineer Antonie van Leeuwenhoek (1632–1723) needed to observe **microorganisms**.

A lens is a specially shaped piece of glass that alters the path of the light rays passing through it. The first recorded lenses were used by Roman inventor Lactantius (c. 240–320) to focus the Sun's rays and to light fires. He made the devices by filling glass globes with water.

Around 1660 English scientist Robert Hooke (1635–1703) built the

Key Components

The compound microscope

Compound microscopes are the standard tools used in laboratories for viewing small details or microorganisms. The specimen is placed on a glass slide and usually covered with a thin glass cover slip before being positioned on the viewing platform. Light from the light source is reflected into the condenser lens using an adjustable mirror. The condenser lens bends the light through a hole in the viewing platform and concentrates it onto the specimen. Light rays that pass through the specimen are directed up the lens tube by the objective lens, which also magnifies the image. The eyepiece lens, at the top of the lens tube, adds more magnification and directs the light into the viewer's eye. Turning the focus adjustment knob moves the entire lens tube up or down, bringing the image into focus.

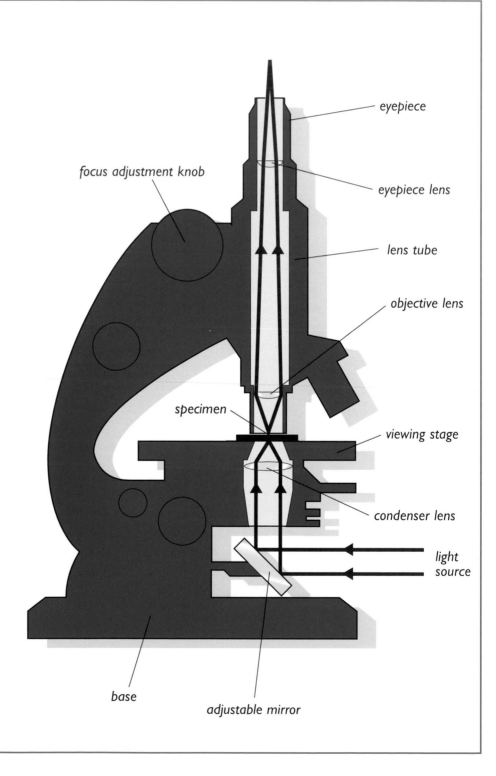

focus adjustment knob

eyepiece

eyepiece lens

lens tube

objective lens

specimen

viewing stage

condenser lens

light source

base

adjustable mirror

first compound microscope. In this type of microscope two lenses or sets of lenses are used. One, the objective, is of small diameter and is positioned near to the object of study. The viewer looks into a second lens, called the eyepiece, which is much larger and is used to further magnify the image produced by the objective.

Telescopes

German-born Dutch eyeglass maker Hans Lippershey (c. 1570–1619) probably invented the first telescope in 1608. According to legend, it was discovered by chance when children playing in his workshop accidentally lined a pair of lenses up with the church tower. Telescopes have a simpler design than compound microscopes, because the light rays they collect are from farther away and traveling in almost parallel lines.

As soon as astronomers began to use telescopes to look at the sky (around 1609), they realized that another factor, called light grasp, influences the performance of instruments of this type. A biologist

▲ *One of the Keck Telescopes at the Mauna Kea observatory in Hawaii. These are the world's largest optical telescopes.*

◄ *A selection of instruments, including a telescope, a MAGNETIC COMPASS, and a PENDULUM CLOCK, that belonged to Galileo Galilei. Galileo was an important 17th-century scientist, making discoveries in the fields of physics, astronomy, and mathematics.*

can simply switch on a lamp and throw more light on a microscope specimen, but an astronomer cannot do the same for a distant planet or star. The very first telescopes used very small lenses to gather light, but astronomers such as Galileo Galilei (1564–1642) in Italy soon realized that the larger the lens, the more light it could capture. From then on lens telescopes became bigger and bigger.

Unfortunately, there is a serious problem with using larger lenses—as they get bigger, they also have to get thicker. Thick glass soon starts to block out the light that astronomers are so eager to collect. In 1663 Scottish astronomer James Gregory (1638–1675) found the solution. Since all incoming light from space arrives in parallel lines, why not simply reflect it to a focus with a curved mirror, rather than bending it with lenses? Mirror telescopes can be made larger than lens telescopes, and they have revolutionized astronomy. The largest steerable mirror telescopes today are the twin 30-ft (10-m) Keck Telescopes on Hawaii.

Scientific Principles

Electromagnetic spectrum

Light has the same properties as other waves—specifically, wavelength, the distance between two peaks or two troughs of a wave, and **frequency**, the number of waves that pass by a fixed point in a second (**1**). For centuries these properties could not be measured, but the fact that lenses could bend light was one demonstration that it could be thought of as a wave. Another was the fact that a wedge-shaped glass prism could split white light apart into a spectrum of colors. This is because the bending effect of the glass is related to the wavelength of the light passing through it—red light, with the longest wavelengths, is bent more than blue light. Today, we know that visible light wavelengths range from red, at 740 nanometers (nm; billionths of a meter), down to violet, at 390 nm.

In fact the **electromagnetic** spectrum extends far beyond visible light on both sides (**2**). In 1800 British astronomer Sir William Herschel (1738–1822), using a thermometer to measure the temperatures of different colors of sunlight split through a prism, discovered that the hottest zone was actually beyond the red end of the spectrum in the region now called the **infrared**. Since then many other types of **radiation** have been discovered, and the forms of electromagnetic radiation that are used by people extend from **radio waves** (with the longest wavelengths), through infrared and visible light, to **ultraviolet**, **x-rays**, and **gamma rays** (with the shortest wavelengths). As the wavelength of these forms of radiation gets smaller, the frequency of waves gets higher, allowing them to carry more energy.

In theory the electromagnetic spectrum has no limits at either end, but natural sources of radio waves and gamma rays get much rarer at the longest and shortest wavelengths.

1 — wavelength / fixed point / light wave

2

WAVELENGTH		EXAMPLES
0.003 nm	gamma rays	nuclear explosion (gamma rays)
0.03 nm	x-rays	x-ray picture
10 nm	ultraviolet	
	visible light	the Sun (ultraviolet to infrared)
	violet indigo blue green yellow orange red	
	infrared	electric heater (infrared)
0.1 mm	microwaves	micro-wave oven
0.8 m		
	radio waves	radio and television
3,000 km		

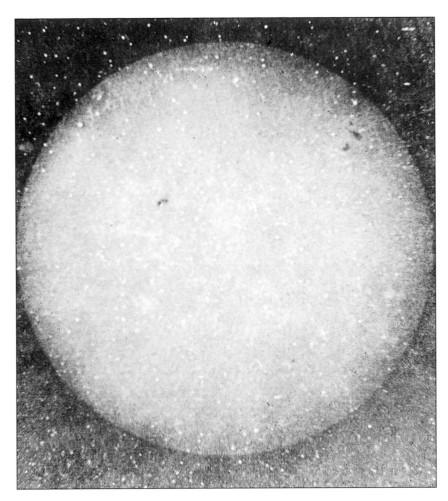

▲ The first photograph of the Sun, taken by Foucault and Fizeau on April 2, 1845. Photography was important in the development of modern science because it took personal observation out of the recording process and produced largely unbiased scientific data for the first time.

Measuring light

Optical instruments such as telescopes and microscopes can magnify images and concentrate light, but for centuries they were only aids to the eye. Since the discovery that visible light is just one small part of a much larger electromagnetic spectrum, however, scientists have realized that the electromagnetic radiation given off or reflected by objects can provide much more information than just how things look.

In order to get this information, scientists had to develop more accurate methods of measuring light. Photographic film was the earliest method of recording observations. As early as 1840 British instrument maker John Dancer (1812–1887) was taking PHOTOGRAPHS of microscopic objects, while in 1845 French physicists Jean Foucault (1819–1868) and Armand Fizeau (1819–1896) took the first photographs of the Sun through a telescope.

Photography could also show things that scientists themselves could not record. As the sensitivity of photographic films improved, exposure times could be reduced to small fractions of a second. It was then possible to study physical and chemical processes occurring over short time intervals. Most famously, British-born U.S. photographer Eadweard Muybridge (1830–1904) used rapid exposures to study the motion of animals in the 1870s.

Perhaps the most important discovery was that photographic film could detect radiation invisible to the human eye. In 1895 German-born Dutch physicist Wilhelm Röntgen (1845–1923) recorded X-RAYS on film, and just a year later French physicist Antoine-Henri Becquerel (1852–1908) accidentally discovered radioactivity when a **radioactive** source fogged a photographic plate. In the late 19th century chemists also developed photographic chemicals that could detect infrared and ultraviolet (UV) light.

Spectroscopy

Armed with films that could detect different wavelengths of radiation and capture fine detail for later measurement, scientists began to study the spectrums produced when light from objects is split through a prism. This technique, called spectroscopy, became an important new field of science. As early as 1814 German physicist Joseph von Fraunhofer (1787–1826) had built the first spectroscopes, but he had no means of recording the spectra they produced. By splitting the light produced by different sources, he discovered that different elements produce light of different wavelengths. Fraunhofer discovered a large number of dark lines at fixed positions in the Sun's spectrum and figured out that they were produced by chemical elements in the Sun absorbing light at these wavelengths. Photographic film allowed the lines, called Fraunhofer lines in his honor, to be measured accurately for the first time, and it was then possible to determine the chemical composition of the Sun's atmosphere. Today, spectroscopy has become a valuable tool for looking at the atmospheres of distant stars and understanding the internal structures of elements.

▼ *When white light is shone through a prism, it is split into its component parts. In the visible spectrum these are red, orange, yellow, green, blue, indigo, and violet. An instrument called a spectrometer is used to study the properties of light.*

▲ *A hologram appears to be a random pattern of dots and stripes, but when lit with a laser beam, it produces an image.*

Lasers

All the uses of light mentioned so far have been in passive measuring devices, but light can also be used in active measurement—firing light at an object can be a useful way of investigating it. This often involves a special type of light called laser light.

Laser light is useful because all of its properties are so well defined (the speed of light is also fixed). It can be used in surveying to accurately measure the distance between two points by recording the time taken for a laser pulse to reach a reflector and return. It can also be used for creating holograms—three-dimensional images of objects. Because laser light can be made in very short pulses, holography can capture very short-lived or rapidly changing phenomena. A hologram can be measured in the

Scientific Principles

Laser light

The word laser stands for *light* amplification by stimulated *e*mission of *r*adiation. Normal light is a jumble of different wavelengths, overlapping with each other and out of phase—the peaks and troughs of different waves are out of step with one another. Laser light, on the other hand, is monochromatic (which means all the waves have exactly the same wavelength) and coherent (which means all the peaks and troughs are in step) creating a very intense source of light. A ruby laser contains a rod of synthetic ruby crystal, two mirrors, and a flash lamp for producing visible light. Light from the flash lamp is absorbed by chromium **ions** in the ruby, raising some ions to an excited state from which they emit light. This light is reflected repeatedly through the crystal, stimulating the emission of more light energy in phase with it, and finally forming an intense red beam.

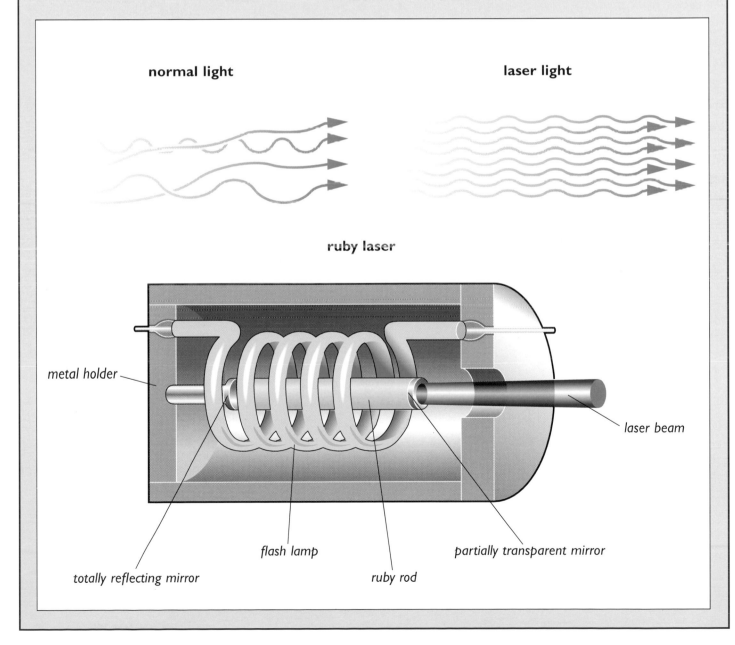

normal light

laser light

ruby laser

metal holder

laser beam

flash lamp

partially transparent mirror

totally reflecting mirror

ruby rod

same way as the real object it records, so it is a very useful tool for scientists.

Measuring temperature

The temperature of any material is actually a measure of the energy of its individual **atoms** and so can give scientists a lot of information. Thermometers were one of the first scientific instruments to be invented, but they took a lot of work to perfect.

A simple thermometer measures the expansion or contraction of a material as it gets hotter or colder. This happens because at higher temperatures the atoms in a material have more energy and move around more quickly. The first thermometer, built by Galileo around 1592, measured the expansion of air. However, air is not an ideal material because it expands very rapidly.

A better answer was to measure the slower expansion of liquid in a tube, with a reservoir of the liquid in a bulb at the base of the tube. Various liquids were tried until German physicist Daniel Gabriel Fahrenheit (1686–1736) popularized the use of the liquid metal mercury in the 18th century. Mercury has a low freezing point and a high boiling point, and it expands only slightly as the temperature changes, so mercury thermometers could be small and still cover a wide range of temperatures.

More recently scientists have found other properties that change steadily with temperature. Bimetallic strip thermometers use a strip made from two metals that expand at different rates. As the temperature changes, the strip bends, and this movement can be transferred to a display dial. The strip can also be designed to close an electrical switch at a certain temperature—as in a CENTRAL HEATING thermostat, for example.

The most accurate thermometers used in laboratories today are

◀ A type of digital thermometer often used in scientific experiments. This picture shows that the temperature of the melting ice in the flask is 0°C, which is equal to 32°F.

electrical. One type measures how the electrical resistance of a special RESISTOR changes with temperature. Another type, called a thermocouple, uses the temperature difference between two ends of a circuit to create an ELECTRIC CURRENT.

Temperature scales

As mercury thermometers became widespread, there was more need to agree on a standard temperature scale. Until this time there were many different scales in use, but if scientists wanted to exchange information, they had to employ a common temperature scale. In 1714 Fahrenheit devised a scale that set the freezing point of a water-salt mixture at 0°F, the freezing point of pure water at 32°F, and the boiling point of water at 212°F. The Fahrenheit scale was popular for many years and is still used in the United States, but most other countries have now adopted the simpler Celsius system (invented in 1742 by Swedish astronomer Anders Celsius, 1701–1744), which sets the freezing and boiling points of water at 0°C and 100°C respectively.

Measurement scales are very important in science, and the method used for devising temperature scales is copied in scales for many other properties. The key is to measure a property at two fixed points and create a scale of equal divisions between them. In most temperature scales these fixed points have no real significance—the boiling and freezing points of water are chosen simply because of water's importance to us.

An exception to this general rule is the Kelvin scale, which is based on a really significant temperature called absolute zero. In 1848 British mathematician and physicist William

Thomson (1824–1907), later Baron Kelvin, measured the rate of expansion in a gas and traced it back to the point at which the gas atoms would stop moving around and have no volume. This point was −460°F (−273°C), which is now known as absolute zero. On the Kelvin scale absolute zero becomes 0°K, while a 1°K temperature difference is equal to 1.8°F (1°C).

Laboratory equipment

As experimental science became established in the 18th and 19th centuries, a wide variety of different laboratory equipment was invented for accurate investigation and

▲ *Baron Kelvin, who, in 1848, first proposed the existence of absolute zero— the lowest temperature theoretically possible. The temperature scale based on this value was later named the Kelvin scale in his honor.*

🔑 Key Components

Calorimeters

A typical modern calorimeter consists of a thermometer surrounded by a material that absorbs the heat to be measured, such as water. The thermometer measures the temperature change in the water caused by, for example, burning a match. If the amount of energy needed to raise the water temperature by a certain number of degrees is known, then a chemist can use the calorimeter reading to work out the amount of energy released by the burning match.

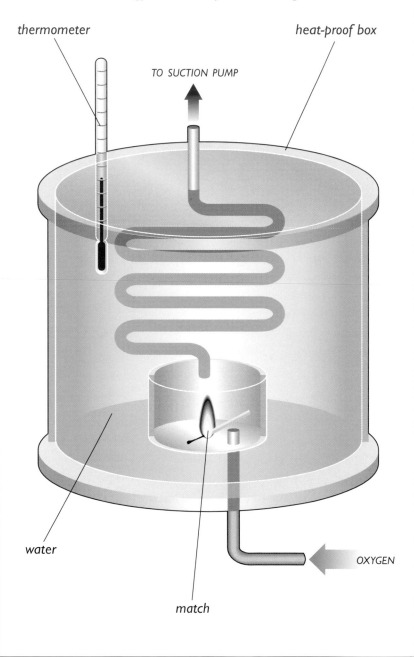

thermometer

heat-proof box

TO SUCTION PUMP

water

match

OXYGEN

measurement. Pioneering scientists began to study effects in more detail, experimenting with atoms, chemical reactions, microorganisms, electric currents, and **magnetic** fields and instruments increasingly provided the only contact with phenomena.

Titration

Titration is a technique for accurately measuring volumes of liquid or gas in chemical reactions, using a device called a burette. A liquid burette is a long, thin tube with a tap at the bottom. Close-spaced divisions up the side of the tube are used to measure the volume of liquid released from the tube. A gas burette works on the same principle but has a reservoir of liquid that keeps the gas under pressure and pushes it out of the tap when it is opened. Along with weighing balances, burettes allowed accurate measurements of the proportions of different chemicals that react with each other—giving the first clues to the proportions of different elements within them.

Calorimetry

In the late 18th century scientists realized that temperature and heat are not the same thing—heat is the energy transferred between objects to make them hotter or cooler. In 1760 French chemist Joseph Black (1728–1799) suggested that the same amount of heat (which he called caloric) could raise the temperature of different materials by different amounts. Two Frenchmen, chemist Antoine Lavoisier (1743–1794) and mathematician and astronomer Pierre Simon Laplace (1749–1827), invented the first heat-measuring device (called a calorimeter) in 1780. Heat energy is measured in units called calories.

Chromatography

Chemists often need to separate the different types of **molecules** in a chemical mixture. Chromatography is the overall name for a series of different separation methods that all involve dissolving the mixture into a liquid solution or gas mixture and separating the different molecules by passing them through a stationary substance, usually a finely divided solid or filter paper. Many different methods can be used to drive this process. The simplest is to rely on the varying weights of the molecules—heavy ones sink to the bottom, while light ones rise to the top. Another method, called electrophoresis, turns the molecules into ions (electrically charged particles) and distributes them through an electric field by the ratio of electric charge to mass.

▲ *French chemist Antoine Lavoisier, one of the founders of modern chemistry, invented a heat-measuring device and discovered and named many chemical elements.*

▶ *Many scientific instruments are attached to space PROBES and SATELLITES to study conditions in space. This satellite carries a type of spectrometer that measures ultraviolet light.*

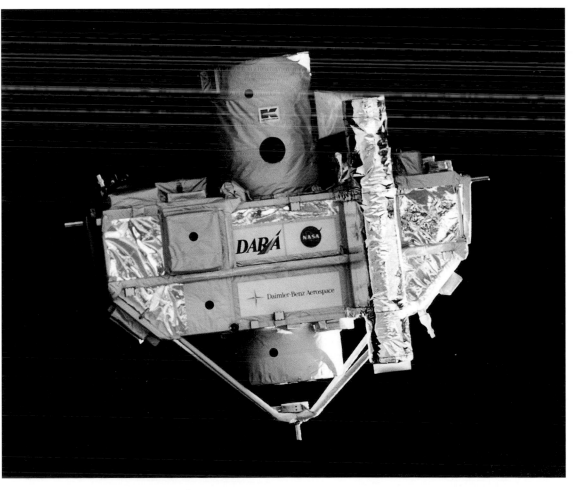

Mass spectrometry

Sometimes chemists need to analyze a sample of material without any clues to what it contains. Often they do this using chemical reactions, but a more advanced method is called mass spectrometry. This involves turning the mixture into a gas and bombarding it with particles to knock **electrons** out of the atoms and produce **ions**. The ions are then accelerated and bent around a curved tube by electric and magnetic fields—the amount by which they are bent depends on the ratio of charge to mass. The arrival of the particles at the end of the tube is registered as an electric current. Since each atom or group has a unique charge-to-mass ratio, it arrives at a different position in the detector and can be identified.

Probing the atom

Improvements in laboratory equipment allowed chemists to understand how molecules are made up from atoms of different elements, but until the start of the 20th century the internal structure of the atom was a mystery. Scientists today use a variety of instruments to study subatomic particles—the building-blocks that make up atoms.

Radiation detectors

The first clues to the structure of the atom came from studying radioactive elements, which give off electrically charged bodies called alpha and beta particles and high-energy gamma rays. Radioactivity was discovered in 1896 by Antoine-Henri Becquerel, who realized that alpha and beta particles are excess subatomic particles (electrons, **neutrons**, and **protons**) being given off by unstable atoms, while gamma rays are a form of energy released as the atom settles back down.

The best-known radiation detector is the Geiger counter, which is filled with atoms of argon gas. Radioactive

▼ *This device is a Geiger counter, which is used for detecting radiation. It was invented by German physicists Hans Wilhelm Geiger and Walther Müller.*

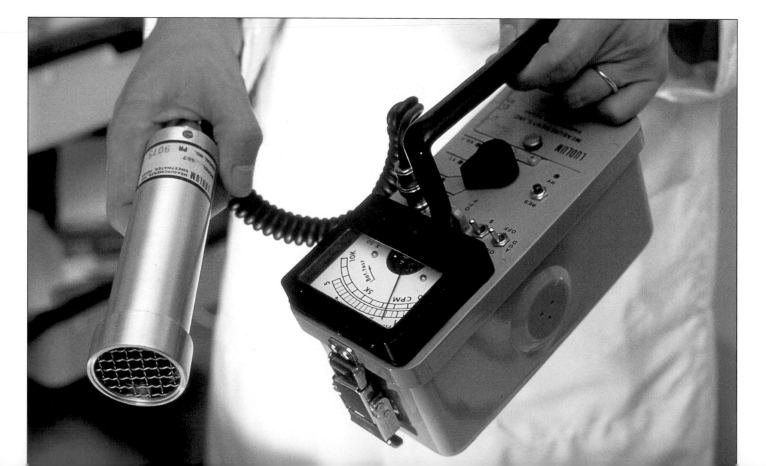

particles passing through the gas collide with these atoms, turning them into ions, which are then accelerated by an electric field and are detected as a burst of electric current.

Electron microscopes

Electrons are subatomic particles that carry a negative charge. They are released from hot metal filaments and can be shaped into a beam by **electrodes**. This beam has many uses—it is involved in producing the picture on a television screen, for example—but it can also be used to study materials at an atomic scale. This is because in a microscope a beam of electrons can be treated in much the same way as a beam of light. There are two main types of electron microscope: transmission electron microscopes and scanning electron microscopes (*see page 36*).

Particle accelerators

Protons, neutrons, and electrons are the most common of many subatomic particles—most of which cannot exist on their own under normal conditions and are usually locked up deep within larger particles. Particle accelerators are instruments that use magnetic and electric fields to send charged subatomic particles around and around a huge circular ring before smashing them into each other and studying the smaller particles that are released for just a fraction of a second. Since U.S. physicist Ernest Orlando Lawrence (1901–1958) built

▲ *A group of tiny mites living on a wasp's knee. This clear, three-dimensional image was created with a scanning electron microscope.*

Scientific Principles

Electron microscopes

A transmission electron microscope fires an electron beam through a thin sample of material. The electrons that pass through light up a screen, creating a picture. The wavelength of an electron depends on its speed. Using electric fields, it is not difficult to accelerate electrons to the speeds required for wavelengths 1,000 times shorter than that of visible light, so they can be used to show details 1,000 times smaller than those seen through an optical microscope.

Scanning electron microscopes have lower power than transmission electron microscopes but can look at solid objects, rather than just thin slices. The electron beam is scanned across the surface, and the intensity of the electrons scattered backward from the object is used to build up an image.

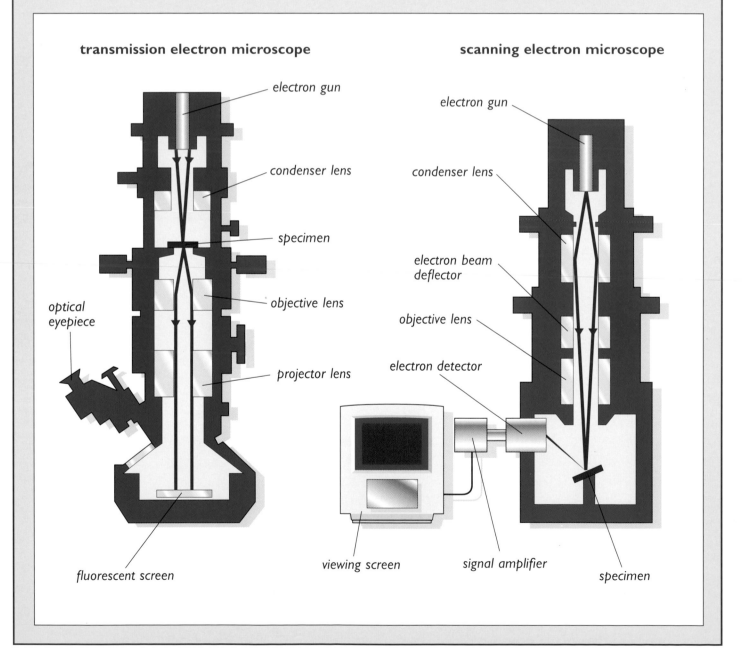

transmission electron microscope

electron gun

condenser lens

specimen

optical eyepiece

objective lens

projector lens

fluorescent screen

scanning electron microscope

electron gun

condenser lens

electron beam deflector

objective lens

electron detector

viewing screen

signal amplifier

specimen

Facts & Figures

● Particle accelerators are able to bring atomic particles up to huge speeds and energies. For example, particles in cyclotrons can reach energies of 25 million electronvolts (MeV).

● The most powerful particle accelerator in the world is the Tevatron at the Fermi National Accelerator Laboratory (Fermilab) in Illinois. Particles in this machine, which has a diameter of 1¼ miles (2 km), collide at up to a thousand billion electronvolts (TeV).

● The largest accelerator in the world is the Large Electron Positron Collider (LEP), which has a circumference of 16.8 miles (27 km), at the CERN laboratory in Switzerland.

the first accelerator of this type in 1931, they have revealed dozens of new particles and brought us much closer to understanding the fundamental laws of nature.

Our world and its weather

Scientists use a wide variety of different instruments to measure the world around us. These range from seismographs, which measure the intensity of earthquakes, to

▼ *Part of the circular tunnel that houses the Large Electron Positron Collider, the largest particle accelerator in the world.*

barometers, which are designed to measure variations in air pressure and help us predict the weather.

Measuring the Earth

Earth is not a stable planet—as the rocky plates that form its surface scrape past one another, they often make the ground shake. The earliest device used to measure these events, called earthquakes, was built by Chinese scientist Zhang Heng (A.D. 78–139) in A.D. 132. It consisted of a set of metal balls delicately balanced above eight carved dragons arranged in a circle. When an earthquake shook the ground, the balls dropped down out of the mouths of the dragons; the number of balls that fell showed the strength of the earthquake, while the dragon whose mouth they fell out of showed the direction of the earthquake's epicenter (the point on the Earth's surface directly above the point at which the earthquake began). Today geologists measure earthquakes with instruments called seismographs.

Measuring wind and rain

A device used for measuring wind speed is called an anemometer, and the most common type relies on the wind to turn some kind of spinning flat surface called a vane. The best-known type of anemometer uses cups on the ends of four arms instead of vanes and was created by Thomas Robinson in 1846. The spinning cups turn an axle, and the axle's rate of rotation is recorded on graph. Many

▶ *A revolving-cup electric anemometer, designed to measure wind speed. This anemometer is located on top of the Jodrell Bank radio observatory near Manchester, England.*

Key Components

Seismographs

A horizontal seismograph is a simple device with a pen hung from a pivot and weighted near its tip so it always hangs straight down. A paper graph rolls slowly underneath it, and the pen draws a straight line. When the ground moves in an earthquake and produces horizontal motion, the pen remains in a vertical position, but the graph moves underneath it, so the lines become jagged. A vertical seismograph works in much the same way, but it detects vertical motion. Earthquake strengths can be worked out from the amount of disturbance on the resulting graphs, which are called seismograms. Using seismograms from different locations, scientists can calculate the position of the earthquake's epicenter.

horizontal seismograph

vertical seismograph

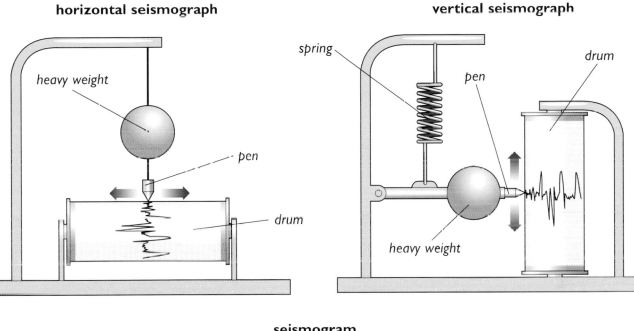

seismogram

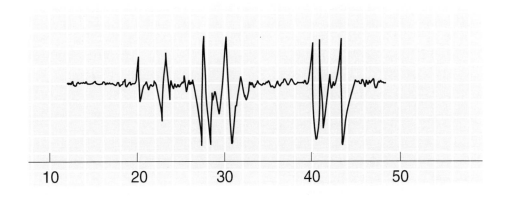

Society and Inventions

Predicting earthquakes

Being involved in an earthquake is a terrifying experience, and many people have died, been seriously injured, or lost their homes and possessions in these natural disasters. Earthquakes are impossible to predict at the moment, but geologists hope they may be able to get advance warning by studying the movements of fault zones— places where the Earth's plates meet. At the San Andreas Fault in California geologists are using lasers to measure the distance between the two sides of the fault. By firing a laser at a reflector on the other side of the fault and measuring the time it takes to return to a detector, the distance between the laser and the reflector can be calculated to one part in ten million.

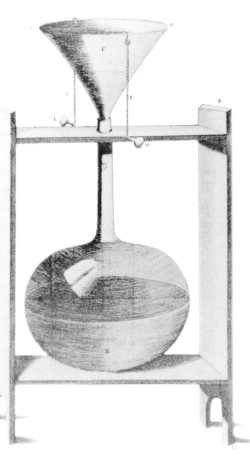

▶ *This rain gauge, constructed by Robert Hooke in 1695, was used to measure the rainfall in Oxford, England, each week between August 1695 and August 1696.*

weather instruments automatically record their information—usually on a paper graph wrapped around a slowly turning clockwork drum, with drawing pens mechanically linked to the instrument itself.

The rain gauge is probably the simplest of all weather instruments and was invented by Robert Towneley in 1677. It consists of a wide funnel (to increase the collecting area) emptying into a transparent tube, which collects the rainwater. A scale up the side of the tube indicates how many inches or millimeters of rain have been collected (taking into account the wide mouth of the funnel). Automatic rain gauges have a recording pen linked to a float on top of the water. The collecting tube is emptied out each day with an automatic siphon.

Pressure measurement

Changes in air pressure have a major effect on the weather and are measured by barometers. Although we cannot feel it, the atmosphere above us is pressing down with the force of 14.7 lb per square inch (one kg per square cm). Italian physicist Evangelista Torricelli (1608–1647) discovered atmospheric pressure in the 1640s and built the first barometer. It consisted of a tube sealed at the top end, filled with mercury, and upended in a bowl containing more mercury. Although the column of mercury inside the tube is pulled downward by **gravity**, the pressure of air pushing down on the mercury in the bowl supports it. The height of the supported column depends on the surrounding air pressure but is usually around 30 in (76 cm). Torricelli used mercury in his barometer because he realized that a similar column of water (which is much less dense than mercury) would be over 30 ft (9 m) tall.

The first barometers were read simply by looking at a scale alongside the glass tube of mercury, but in 1665 English scientist Robert Hooke added a circular dial to the top of the barometer. A wire running to a float on top of the mercury column turned the needle of the dial, indicating the pressure.

Another type of pressure-measuring device is the aneroid barometer, invented by Lucien Vide in 1844, which contains a sealed metal cylinder that has had air pumped out of it. The low pressure inside the cylinder makes it expand and contract depending on the surrounding air pressure, and these changes are recorded on a pressure dial. Aneroid barometers are often used in aircraft

because they are more compact than mercury barometers.

Heat, humidity, and radiation

When meteorologists want to measure the temperature of the air, they use the same mercury or electrical thermometers as other scientists. An air thermometer has to be placed in the shade, however, because direct sunlight heats metal much more quickly than air, and so the thermometer could end up

▲ *Italian physicist Evangelista Torricelli creates the first barometer by carefully lowering a mercury-filled glass tube into a bowl containing more mercury.*

41

measuring the temperature of its own bulb rather than the air around it. A special weather station housing called a Stevenson screen (a white screen with louvered slats) allows free circulation of air while keeping out sunlight and is used for thermometers and barometers, which can both be affected by heat from sunlight.

The Stevenson screen is also a useful housing for hygrometers—devices used to measure the moisture in the air, or humidity. The first hygrometer, invented by Italian artist and inventor Leonardo da Vinci (1452–1519) around 1500, measured the weight of a ball of wool as it absorbed and released water. Hygrometers today are based on

measuring the temperature of two different thermometers, one of which is surrounded by wet cotton wool. The rate at which water from the cotton wool evaporates depends on the amount of moisture in the surrounding air, and the evaporation has a cooling effect on the thermometer bulb. So the difference in temperature between the two thermometers is largest when the air humidity is low and the rate of evaporation is high.

Many weather instruments have to be kept out of direct sunlight, but meteorologists still need to collect and measure the power of sunlight. Instruments to measure the Sun's radiation are called pyrometers. They

▼ *A weather station in Finland. Instruments for measuring temperature, pressure, and humidity are housed inside a Stevenson screen, which shields them from sunlight. The slats of the screen are louvered (horizontal and overlapping).*

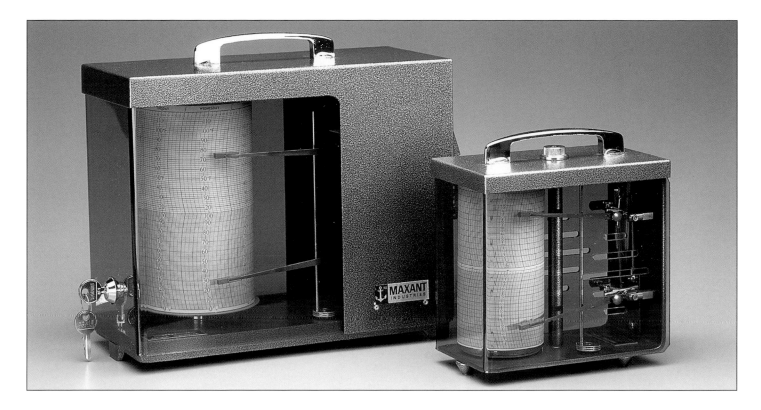

have a glass dome that isolates them from the surrounding air temperature and focuses sunlight onto a thermopile—a series of temperature-sensitive electrical circuits called thermocouples. The voltage produced by the thermopile is generated purely by radiation from the Sun.

Weather stations, buoys, balloons, and satellites

To compile a weather map and make a forecast, **meteorologists** need to collect many different types of information from a lot of different points, on land, at sea, and in the air. On land a full set of instruments is housed in a protective weather station. Some weather stations are still checked daily or weekly by hand, but others are automatic—they record their data electronically and transmit it to a collecting center.

Weather buoys are automatic floating weather stations in the middle of the oceans, while weather balloons, or RADIOSONDES, carry

instruments high into the atmosphere to measure conditions there. Both buoys and balloons can transmit their data back to weather centers via **satellites** high above the Earth. **Remote-sensing** weather satellites carry powerful cameras that can photograph whole weather systems or focus on small details. To gain the maximum amount of information, they use cameras that can detect many different wavelengths of light and even use RADAR (radio waves bounced off the Earth's surface) to show rainfall patterns, wave heights, and wind strengths. Since the arrival of satellites, meteorologists have been able to see large-scale pictures of the world's weather.

See also:
➤ The Age of Electricity (Vol. 10)
➤ Exploring Other Worlds (Vol. 1)
➤ Measuring Time (Vol. 5)
➤ Medicine Meets Science (Vol. 8)
➤ Modern Power (Vol. 10)
➤ Photos and Flicks (Vol. 3)

Technology and Measurement

How measurement makes use of advanced technology, including ultrasound, telemetry, and biosensors

◀ *Space* PROBES *are perhaps one of the most advanced uses of technology in measurement. This is an artist's impression of the three parts of the Mars Surveyor probe: the orbiter (top), the lander, and the* SOJOUNER ROVER *(far right). The lander collected data from the Martian surface and relayed it back to Earth via the orbiter using the principles of* TELEMETRY.

Measurement often seems very straightforward. What could be simpler than measuring your height or telling the time? Despite its apparent simplicity, however, measurement sometimes presents scientists with very complex problems. How, for example, is it possible to measure accurately the age of rocks or dinosaur bones? How can measurements be made in dangerous places, such as the core of a NUCLEAR REACTOR or the murky depths of the seabed? Problems such as these have tested the ingenuity of inventors to their very limits.

▶ *French physicist Pierre Curie showed that electricity could be generated from certain crystals, such as quartz, by the application of pressure.*

Measuring with sound

Bats, porpoises, and some other creatures find their way around by sending out a beam of high-**frequency** noise and listening for reflections from obstacles they need to avoid or creatures they can eat. Echolocation, as this is called, was first described by Italian scientist Lazzaro Spallanzani (1729–1799). His discovery has inspired a method of measurement and detection using high-frequency ULTRASOUND.

Ultrasound has been studied by several inventors. In 1846 British physicist James Joule (1818–1889) demonstrated the so-called "magnetostrictive" effect. He found that a bar of a **magnetic** material, such as iron, changes its size when placed in a magnetic field. Using this principle, ultrasound can be created using a high-frequency electric current to magnetize a metal rod.

An alternative method of generating ultrasound waves is called **piezoelectricity**, which means "electricity from pressure." This was first demonstrated in 1880 by French physicist Pierre Curie (1859–1906), who showed that when certain crystals, such as quartz, are squeezed, they generate an electric current. Conversely, when an electric current is applied to the crystals, they vibrate and emit ULTRASOUND waves. O. Millhauser pioneered ultrasonic testing of flaws in metal in 1933.

During **World War I** (1914–1918) ultrasound was used as the basis of the marine navigation and detection system sonar (*so*und *na*vigation *r*anging) following the work of French physicist Paul Langevin (1872–1946). British meteorologist Lewis Richardson (1881–1953) developed echo sounding in ships in the 1930s. Medical uses of ultrasound include scanning for tumors and checking fetuses in the womb. These techniques were pioneered by British physicians Ian Donald (1910–1987) and Stuart Campbell (born 1936).

Scientific Principles

Ultrasound and ultrasonics

Ultrasound is high-frequency sound beyond the range of human hearing—typically 20,000 **hertz** (Hz) or higher. Ultrasonics is the name given to the various uses of ultrasound, which include detecting flaws in metals, checking the human body for tumors, and surveying the seabed.

Most of these applications involve emitting a beam of ultrasound and seeing how this beam changes when it is reflected back by objects. Echo sounding, or sonar, was one of the first and the simplest uses. If a ship sends out a beam of ultrasound toward the seabed (*below*), the depth of the ocean at any given point can be measured by timing how long it takes for the beam to be reflected back.

Ultrasonics has found many applications in industry, particularly in what is called nondestructive testing. This involves subjecting materials to beams of ultrasound to detect cracks and other flaws that are invisible to the naked eye. The cracks show up as echoes in the ultrasound beam. Aircraft ENGINES, wings, and other structures can be routinely tested in this way without taking them out of service.

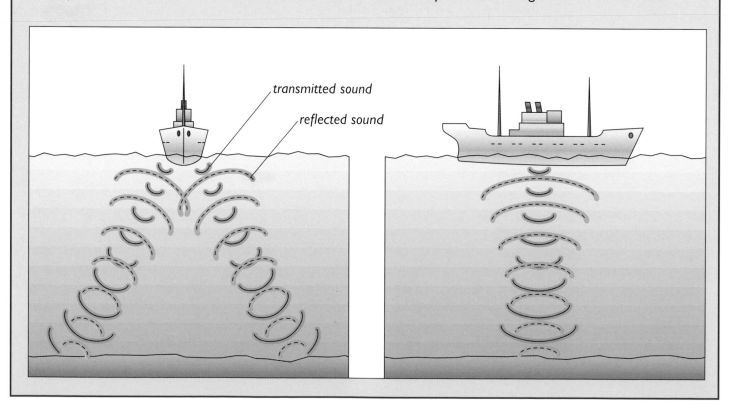

transmitted sound

reflected sound

▲ *A satellite image of New York City taken by Landsat 5. Buildings appear in blue, bare ground in yellow, green vegetation in red, and water in black.*

Measuring at a distance

Sometimes measurements need to be made in remote or even dangerous environments using a technique called telemetry, which means "measuring at a distance." Telemetry was first used in weather forecasting. In 1874 weather measurements made on Mont Blanc, a mountain in France, were transmitted 300 miles (480 km) to a receiving station, where they were decoded and interpreted. By the 1930s this idea had evolved into the modern-day weather BALLOON, or RADIOSONDE. It is a small, unpiloted rubber balloon that rises to an altitude of 20–30 miles (32–48 km) and then bursts. During its ascent it makes regular measurements of air pressure, temperature, and humidity (the amount of water vapor in the air) and sends them by RADIO to a ground station. The measurements from many such devices around the world are collated and fed into COMPUTERS to produce weather forecasts.

Telemetry is now used to monitor and control a wide variety of equipment, including space ROCKETS, the reactors in nuclear power stations, and facilities such as pipelines in remote areas. Telemetry also forms

▼ *A weather balloon, containing instruments for measuring temperature and humidity is about to be released by a meteorologist. This particular model was designed by inventor Jim Scoggins and is named the Jimsphere for him.*

Key Components

Telemetry

All telemetry systems are made up of the same basic components. Most importantly, they need one or more sensors to make measurements at the remote location, such as CAMERAS, pressure gauges, or thermometers (1), as well as a device called a TRANSDUCER for encoding this information into a transmission signal (2). Telemetry systems also need a communication link from the remote location to the base station (3). This is typically a radio transmitter and receiver on, for example, a satellite in orbit and a satellite dish on the ground, but the link can also be a fixed wire that transmits the signal as an electrical current. The communication link may be one way (the remote device simply sends its measurements at intervals back to the base station) or two way (the base station sends instructions on how or when the measurements are to be taken). The final component of a telemetry system is some method of recording (4) and then analyzing (5) the data back at the base station.

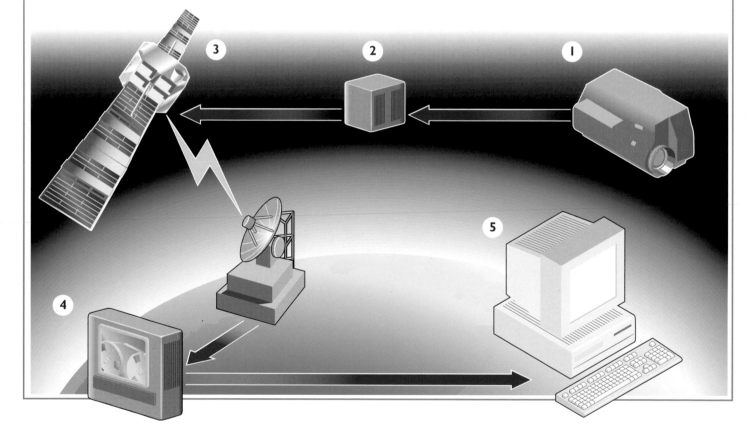

the basis of SATELLITES that sense and measure activities on Earth and other planets. The detailed images that these satellites produce are used for scientific research, mineral exploration, mapmaking, and many other applications. The latest working U.S. satellite, called *Landsat 5,* is fitted with sensitive instruments that measure the different bands of **electromagnetic radiation**.

Looking back before time

Geologists and archeologists often need to date human artifacts or rocks that predate the invention of time

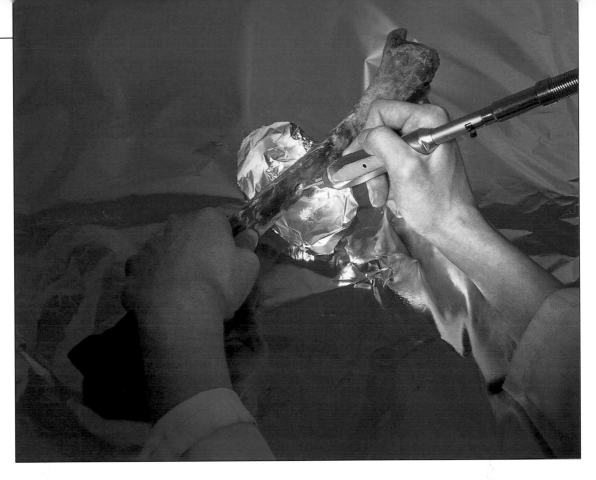

measurements and record keeping itself. Most people know that a tree can be dated roughly by counting the rings inside its trunk. But how is it possible to date a lump of bone that gives no outward signs of its age?

The ingenious solution was developed in 1947 by U.S. chemist Willard Frank Libby (1908–1980), resulting in his award of the Nobel Prize for Chemistry in 1960. His discovery was that all living organisms contain a certain proportion of a **radioactive** type of carbon called carbon-14, or radiocarbon. Radiocarbon very, very slowly decays into a stable type of carbon (carbon 12) at a precise rate. By measuring the amount of carbon 14 left in a bone sample or artifact, it is possible to calculate the age of the material fairly accurately. Radiocarbon dating, as this technique is called, is used to determine the age of once-living organisms. Another technique, called argon-potassium dating, is used for nonliving samples such as rocks.

Measuring electricity

Engineers need to be able to make accurate measurements of current, electric charge, and VOLTAGE, and utility companies need to know how

Facts & Figures

● The first ever artificial satellite—*Sputnik 1*—was launched on October 4, 1957, by the former USSR.

● Today more than 3,000 satellites are orbiting Earth. Most of these satellites were launched from Russia, the United States, Japan, and Europe.

● From its launch in 1972 until its "retirement" in 1978 the U.S. *Landsat 1* satellite generated over 300,000 images of Earth.

How old is the Turin Shroud?

The Holy Shroud of Turin is a cloth bearing a ghostly image of a bearded figure that many people believe is a picture of Jesus Christ. Since 1578 it has been locked in a casket in the Cathedral of San Giovanni Battista in Turin, Italy. But controversy has surrounded the shroud since it first emerged in 1354. Was it the shroud that had clothed Jesus Christ after his crucifixion or just a medieval fake?

The technique of radiocarbon dating allowed the mystery to be settled once and for all. In 1988 scientists from Britain, Switzerland, and the United States each took a small sample of the shroud for dating. It turned out that the Holy Shroud of Turin dates from somewhere between 1260 and 1390 (the **Middle Ages**). Although the Turin Shroud does not date from the crucifixion, it is still revered as a holy object by some members of the Catholic Church.

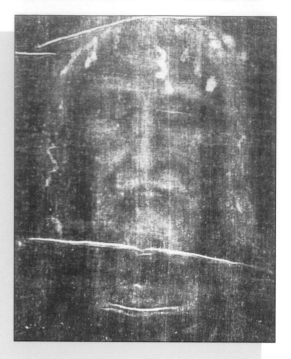

much electrical energy their customers use to bill them accurately.

Early methods of electrical measurement relied on the force of attraction between bodies that had an electric charge. In 1786 British physicist Reverend Abraham Bennet (1750–1799) invented an electricity-measuring device called a gold-leaf electroscope. In this device a piece of

▶ *Charles-Augustin de Coulomb with his invention for measuring electrostatic charge—the torsion balance electrometer.*

gold leaf is attached to an electrical **conductor** and rises upward by an amount that depends on the charge carried by the conductor. French physicist Charles-Augustin de Coulomb (1736–1806) produced a more accurate device called a torsion balance electrometer, which also measured charges using the electrostatic forces it gave rise to.

Devices for measuring electric current relied on the way in which a current could give rise to magnetic effects, which was demonstrated in 1820 by Danish physicist Hans Christian Ørsted (1777–1851). If a piece of wire carrying an electric current is placed in a magnetic field, it is subject to an electromagnetic force that makes it move; the bigger the force, the greater the movement. This simple idea forms the basis of the GALVANOMETER, which is the general name for meters such as ammeters and voltmeters that ultimately measure an electric current. They work much like electric motors, using a needle attached to a coil of iron that pivots in a magnetic field. The first such device was made by French physicist Claude-Servain Pouillet (1791–1868) in 1837 and further developed by French physicist Jacques d'Arsonval (1851–1940). The Western Electrical Instrument Company, based in New Jersey, was the first company to manufacture galvanometers on a large scale.

Electric utilities need to measure not the size of the electric current passing through someone's home but the amount of electrical energy they use in total. U.S. inventor Thomas Edison (1847–1931) developed crude electricity meters that used the chemical effect of an electrical current to indicate the amount of power used. British physicist William Edward Ayrton (1847–1908) and his partner, British electrical engineer John Perry (1850–1920), devised a more accurate instrument. If a magnetic coil is placed near a PENDULUM CLOCK, the coil slows down the clock by an amount that depends on the amount of electrical current flowing through it. A second clock placed away from the coil can be used as a reference so that

the time difference between the two clocks gives a measure of the electricity consumed. Modern electricity meters were devised in 1882 by British-born U.S. engineer Elihu Thomson (1853–1937).

Biological detectors

Despite the long history of measurement, inventors never cease to devise ingenious ways to sense,

▲ Elihu Thomson— inventor of the modern electricity meter—working in his office. Thomson went on to found the Thomson-Houston Company in 1879, which later merged with Thomas Edison's company to become today's General Electric Corporation.

Electricity meters

The heart of modern electricity watt-hour meters is a small electric motor. The rotating part (known as a rotor) of this motor is the familiar spinning disk you see when you look at the meter. The disk is set up so that it spins between the poles of two permanent magnets, which act like a brake. The disk spins around at a speed that depends on the total amount of electricity being used at any given time. The number of revolutions of the disk gives an indication of the total amount of

electricity consumed, to an accuracy of about 0.5 percent. The meter is connected to a circuit breaker, which breaks the circuit if the current exceeds a certain level. This guards against the possibility of overheating and a fire.

In most meters the disk is connected by a system of GEARS to a set of pointers or a **digital** dial that allows the meter to be read. Some meters are able to send their readings automatically to the utility using radio telemetry. In the future maybe all meters will work this way.

◄ *A computer graphic image of a biosensor used to detect trace amounts of copper. The mass of white and brown atoms in the center of the sphere represents the enzyme superoxide dismutase, which reacts with copper atoms (shown in yellow) to change their color to blue. When scientists see this color in the biosensor, they can confirm the presence of copper atoms.*

detect, and measure things. One of the most promising areas of research involves developing "living sensors," using a combination of proteins called **enzymes** and traditional ELECTRONIC COMPONENTS that can detect a chemical or biological change and convert it into an electrical signal. Biosensors, as these devices are called, can be used to measure minute concentrations of certain chemicals.

They are useful for detecting pregnancy (*see box on page 53*), gas leaks, drugs or explosives in luggage, or dangerous pollution in water ways. A "living **electrode**" based on enzymes was developed by British researchers in 1967.

Biosensors are just one example of the field of BIOTECHNOLOGY, which covers everything from baking bread to genetically modifying organisms.

Pregnancy-testing kits

Enzymes are the active ingredients in pregnancy-testing kits, where they are used to measure the amount of glucose in a woman's urine. The enzymes are impregnated on a piece of filter paper, which is dipped into a sample of the urine. If glucose is present, a chemical reaction takes place, and the filter paper changes color, indicating pregnancy. Enzymes are also used to test for the presence of a substance called human chorionic gonadotrophin (HCG), which is another indication of pregnancy.

Enzyme-based sensors are cheap, quick, reasonably accurate, and reliable. Future applications are expected to include helping diabetics regulate their blood-sugar levels and providing early warnings of breast cancer.

Although biotechnology is almost as old as baking, brewing, and FARMING, modern biotechnology begins with the discovery of techniques for adding new genes to existing organisms to produce desired enzymes or other materials or characteristics.

See also:
➤ Exploring Other Worlds (Vol. 1)
➤ Diagnosis and Testing (Vol. 8)
➤ Industrialization and Automation (Vol. 7) ➤ Measuring Time (Vol. 5)
➤ Scientific Investigation (Vol. 5)
➤ The Space Age (Vol. 1)

Society and Inventions

DNA fingerprinting

DNA fingerprinting (*right*) is another ingenious area of measurement and detection based on biotechnology. It has been routinely used by police forces worldwide since it was developed in the 1980s by British biochemist Professor Alec Jeffreys (born 1950). It works on the principle that every person has a unique **genetic** makeup coded in the molecules of DEOXYRIBONUCLEIC ACID (DNA) in their living cells. In other words, a person's DNA is similar to their fingerprints. Along with other evidence, DNA samples taken from a crime scene can be used to eliminate innocent people or help convict suspects.

Surveying, Navigation, and Cartography

How the world is measured, mapped, and traversed

◀ Seafarers have navigated the globe with the help of the compass, rope, and timepiece since the early 13th century, surveying and mapping the Earth's surface as they go.

Surveying is used to mark out points or boundaries on the Earth's surface—or even those above or below the Earth's surface. Navigation is the process of moving from one location to another by design. Cartography, or mapmaking, is common to both.

Surveying is at least as old as large-scale FARMING. As soon as people settled down in large communities to grow crops, marking boundaries became important. In ancient Egypt the Nile River overflowed its banks every year, covering the carefully marked-out fields with mud. The field boundaries had to be fixed anew every year, and so counting out distances and aligning boundaries with known landmarks became essential. Building the great PYRAMIDS at Giza, Egypt, between about 2600 and 2500 B.C. required highly developed surveying skills.

Early navigators

By 1500 B.C. seafarers from the islands of the southwest Pacific were used to traveling on the open ocean beyond sight of land. They must have been able to read clues in the sea and sky. Clouds, winds, and even the color, taste, and smell of seawater all provide signs about location for those who have the skills to interpret them. Most European sailors were much less adventurous than Pacific islanders, and it was not until some 2,700 years later in the 13th century, with the use of a magnetic compass that pointed north, that more than a handful of European sailors risked sailing beyond sight of land.

◀ For centuries people have been fascinated with traveling to the far reaches of the globe, navigating by reference to the Sun.

The Sun and the stars

On sea or land, movements of the Sun and stars indicate direction. The Sun rises in the east and sets in the west, and during the day it moves in a curve across the sky. In the Northern Hemisphere this curve is toward the south; in the Southern Hemisphere it is to the north. So the Sun's position in a clear sky, and its movement over a few hours, can give a rough-and-ready indication of direction.

Viewing the sky from Earth, the stars appear to rotate around the nearest pole (*as above*). A star positioned exactly above the pole would appear to be stationary while all other stars appear to rotate about it. In the Northern Hemisphere the North Star (Polaris) very nearly meets this requirement. In the Southern Hemisphere the stars "revolve" about a point in the Southern Cross (Crux Australis).

Ancient navigators would have used the Sun during the day and the stars at night to navigate by, but these signs alone would have been insufficient. They needed charts or maps, and they needed to take into account all sorts of other signs—winds, cloud patterns, color of the water, and so on. They also used dead reckoning. This involves plotting direction and distance (calculated from the time the ship has been moving and the speed at which it is traveling) onto a map, while making allowance for winds and currents. The ship's progress is estimated, and corrections are made when a reliable navigational feature, such as a headland or harbor, is sighted.

In the 13th and 14th centuries European seafarers plotted their course with compasses, logs, ropes, and sandglasses. The ship's speed was estimated using a piece of wood (the "log") tied to a rope knotted at regular intervals. When the log was tossed overboard and the rope paid out, speed was measured by the number of knots passing through the hand in a sandglass interval of time. Two terms—*logbook,* a daily record of measurements, and *knot,* speed equaling one nautical mile an hour (1.6 km/h)—come from this practice.

The magnetic compass

When the **magnetic** compass was introduced into 12th-century Europe, it revolutionized navigation at sea. Seafarers could now tell the direction of north even when clouds obscured the sky. The first reliably documented use of a magnetic compass for navigation at sea was in 11th-century China. By the end of the 13th century European compasses had a scale—a circle divided into 32 "quarters," or points—for measuring direction, or bearing. A compass could be used for measuring a ship's position by taking bearings from two or more known points, or by taking a bearing and distance from a known location.

Latitude

In the 6th century A.D. a zealous Christian monk overturned the **ancient Greek** and **Roman** idea that the world was a sphere. It was only in the 14th century that Europeans were once again encouraged to accept that the Earth was round rather than flat. Doing so meant that once again parallels of latitude (lines measuring out the distance from the equator north and south to the poles) could be measured. Latitude is a measure of north–south location on the Earth. The equator is located at 0° latitude, while the poles are at 90° latitude. Working out how far (how many degrees you are) from the equator means measuring the position in the sky of the North Star or Southern Cross and the Sun at noon. If you are near the equator in the Northern Hemisphere, the Sun is high in the sky and the North Star is low on the horizon. Near the north pole the North Star is high and the Sun is low.

◄ *A universal compass invented by Nikolaus Rensperger in 1568. Although many seafarers still use compasses to navigate the oceans, most modern vessels are equipped with SATELLITE navigation systems.*

In the 15th century measuring height (altitude) above the horizon involved using an instrument called a cross staff or back staff. By the late 18th century the sextant could measure altitudes with greater accuracy, and this instrument is still used by some sailors today.

Longitude

Longitude is a measure of east–west location on the Earth. Imaginary lines of longitude run from the north pole to the south pole at regular intervals. Longitude and time are directly related. The Earth turns once on its axis (through 360°) in 24 hours. So 15° of longitude is equivalent to one hour. In 1530 Flemish astronomer and scientist Gemma Frisius calculated that the time difference between local noon (when the Sun is at its highest point in the sky) and the noon at a known longitude some distance away gives the longitude (in degrees) between the two. Nowadays the Greenwich Meridian is used as 0°

longitude, and longitude measurements are calculated relative to this line. If your local noon is two hours "behind" Greenwich mean time (GMT), then you are at longitude 30° west; if local noon is one hour "ahead" of GMT you are at longitude 15° east.

One of the biggest problems for sailors in the 16th and 17th centuries was the inaccuracy of their CLOCKS, because the movement of the ship affected pendulum clocks. Without an accurate clock they could not measure the time of local noon properly, and so their measurement of longitude was poor. This could have disastrous consequences. On one occasion in 1707 four British naval ships ran aground and were sunk, resulting in the death of 2,000 mariners. This tragedy was due to a faulty longitude reading.

In 1714 the British government offered a prize of £20,000 (more than $2 million at today's exchange rates) to anyone who could design a reliable

Facts & Figures

● The Straits of Dover between England and France is arguably the busiest shipping lane in the world; over 500 ocean-going vessels may be navigating through the 25-mile (40-km) wide channel at any one time.

● Chicago O'Hare International Airport is one of the busiest airports in the world. It handles over 100 aircraft operations (takeoffs and landings) every single hour.

◀ *A sailor uses a sextant to measure the angle between the Sun and the horizon so he can find out his latitude position.*

and accurate ship's clock, or CHRONOMETER. British watch- and clockmaker John Harrison (1693–1776) made several types of accurate chronometer, including a model successfully tested in 1761. He did not, however, receive his prize money until 1773, when he was 80. The government was reluctant to hand over the money even then and did so only because King George III (1738–1820) intervened. Using an accurate clock in combination with an accurate altitude-measuring device, such as a

◀ British watch- and clockmaker John Harrison, who finally received a prize of £20,000 from the British government for his invention of an accurate ship's chronometer.

 ## Scientific Principles

Gyroscopes

The gyroscope is an instrument that rotates to produce stable direction in space. A simple gyroscope is made up of a spinning ball or wheel (the rotor) and a supporting system of rings known as gimbal rings. Once the rotor is set in motion, the gyroscope resists any attempt to change its direction of rotation. As a result, gyroscopes are often used in flight and navigation instruments. They can provide heading or course information that is unaffected by air turbulence or rough seas.

Large gyroscopes have been used as stabilizers to reduce the rolling of ships at sea. AUTOMATIC PILOTS using gyroscopes can guide a ship or plane closer to a course than a human pilot can. Other gyroscopes are used to guide MISSILES, satellites, space vehicles, and TORPEDOES.

rotational axis

rotor

gimbal rings

▶ Once spinning, the gyroscope's rotor will keep the same rotational axis wherever it is.

Society and Inventions

Modern navigation

The development of electrical equipment, then electronic equipment, and now computers has revolutionized 20th-century navigation. Some of the highlights among these have been the gyrocompass, the GLOBAL POSITIONING SYSTEM (GPS), and radar. The magnetic compass suffers from the disadvantage that the Earth's magnetic field changes over time, and in any case, true north and magnetic north do not correspond. In the early 1900s several investigators in France, Germany, and the United States devised the gyroscopic compass, or gyrocompass. This device contains a flywheel that, once set spinning in a true north direction, does not deviate even when the craft on which it is mounted changes direction. The mechanical gyroscope has since been replaced with modern LASER gyroscopes containing OPTICAL FIBERS. Gyroscopes lie at the heart of many modern guidance systems. An aircraft's autopilot system contains two gyroscopes

▲ *The "artificial horizon" in the cockpit display panel of modern aircraft is based on a horizontally spinning gyroscope.*

set at right angles. Once a course is set, these gyroscopes detect any deviation from the planned course and, via a computer, adjust the aircraft's controls to keep it flying on the intended path.

Today almost any large aircraft or boat can compute its precise position on the planet within a matter of seconds. This is made possible by the Global Positioning System, fully operational since 1995. Twenty-four Navstar satellites orbit the Earth, each broadcasting its exact position and time. At any one time four satellites are above the horizon, and a ship or other surface vessel can detect them with suitable computerized decoding equipment. By cross-referencing the data from three satellites, the GPS decoder can establish the precise position of the craft that is displayed as a longitude and latitude reading accurate to within 50 ft (15 m). An aircraft uses the data from the fourth satellite to fix its altitude. The GPS works under nearly all weather conditions and for most locations on Earth.

Advances in navigation on land and in the air have largely been made possible by the development of radar (*radio detection and ranging; see box on page 61*), pioneered by British and U.S. scientists during **World War II** (1939–1945).

▼ *The orbits of the 24 Navstar satellites used in the Global Positioning System (GPS).*

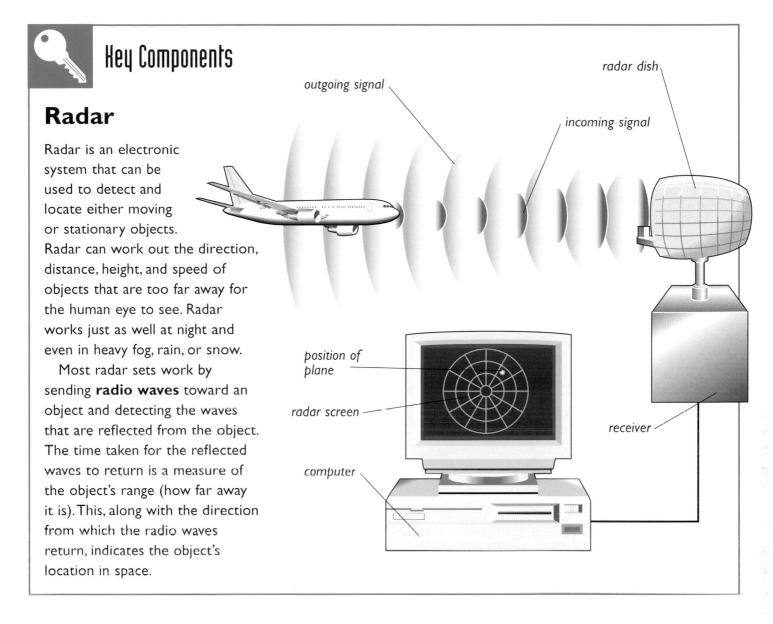

outgoing signal

radar dish

incoming signal

Key Components

Radar

Radar is an electronic system that can be used to detect and locate either moving or stationary objects. Radar can work out the direction, distance, height, and speed of objects that are too far away for the human eye to see. Radar works just as well at night and even in heavy fog, rain, or snow.

Most radar sets work by sending **radio waves** toward an object and detecting the waves that are reflected from the object. The time taken for the reflected waves to return is a measure of the object's range (how far away it is). This, along with the direction from which the radio waves return, indicates the object's location in space.

position of plane

radar screen

computer

receiver

sextant, transformed navigation. By the late 18th century seafarers could use a chronometer, magnetic compass, and sextant to work out their latitude and longitude and so calculate their precise position on the Earth's surface. Nowadays any town or city can be precisely located on the globe or a world map by its latitude and longitude. New York City, for example, is found at latitude 40° north and longitude 74° west.

Charts

In the 6th century B.C. Greek philosopher Anaximander of Miletus (c. 611–547 B.C.) drew the first

known map of the world. His map was based on reports from European explorers and traders. By the 4th century B.C. most Greek and Roman scholars agreed that the world was not flat but was a sphere or globe, or possibly shaped like a cylinder. Around 250 B.C. Greek astronomer Eratosthenes (c. 276–194 B.C.) drew a map of the world that included lines of longitude and latitude (although these lines were in very different locations from the lines used today).

The most famous Greek cartographer (mapmaker) was astronomer and geographer Ptolemy (also called Claudius Ptolemaeus),

who lived in the second century A.D. He was probably the first person to draw a map that took into account that the Earth's surface was curved. His map is extremely inaccurate by today's standards but was a great improvement on what came before. Ptolemy's maps were lost after the fall of Rome in 476 A.D. but were discovered by Portuguese seafarers who visited Constantinople (a city in northwest Turkey now called Istanbul) in 1400. In the intervening period it was Arab cartographers who plotted the best maps.

The usefulness of maps increased dramatically in the late 13th century with the introduction of the graduated magnetic compass. Navigators could now take compass bearings to calculate their position. These measurements were added to charts made from vellum (animal skin), which were called portulans (harbor-finding charts). By the 15th century European sailors were using instruments such as the cross staff, back staff, and astrolabe (an instrument predating the sextant that

was once used to work out the position of planets) to measure latitude, and these measurements too were added to charts. Then, in 1569 a famous Flemish cartographer called Gerardus Mercator (1512–1594) developed a chart that took into account the Earth's curved surface. He treated the Earth as a cylinder orientated north to south. His maps showed landmasses at the equator drawn accurately, but toward the poles the landmasses were increasingly distorted. This did not matter too much because most travel happens in temperate to tropical latitudes, rather than near the poles. His maps were so successful that the Mercator projection, as it is called, is still used in most maps today.

Between the 16th and 19th centuries maps became much more precise as more accurate instruments—particularly clocks, sextants, magnetic compasses, and other surveying and navigational devices—were designed. By the 18th century charts took into account that magnetic north did not coincide with

▲ *A satellite image of the Earth presented as a Mercator projection. On a map using the Mercator projection the lines of longitude and latitude appear as straight lines that cross at right angles. With the aid of a compass these straight lines help navigators plot accurate routes.*

true geographical north. In 1884 the Greenwich Meridian running through Greenwich, London, was internationally accepted as the line of zero longitude.

Modern electronic charts

Since the 1960s **remote-sensing** satellites have been taking photographs from low-level **orbit** in space. These satellite images confirm the work of mapmakers. They also fill in detail for those regions—such as the frozen poles, dry deserts, dense forests, and mountainous regions— where readings from land or low-

flying aircraft have not been made. Nowadays satellites carry an array of sensors that detect the color, temperature, and texture of the land and sea surface. They tell us a great deal about events happening on land and sea and even below the sea surface. As often as not, today's charts are images on screens instead of lines on paper.

See also:
➤ The Birth of Electronic Communications (Vol. 3)
➤ Scientific Investigation (Vol. 5)
➤ The Space Age (Vol. 1)

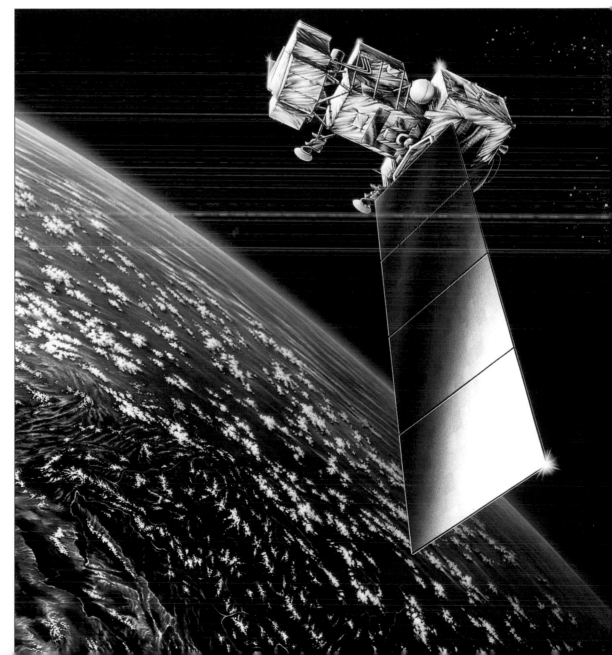

▷ *An artist's impression of the Landsat 6 satellite in orbit around the Earth. This satellite was designed to produce detailed maps of the planet's surface, but one of its motors failed in orbit, and the satellite reentered Earth's atmosphere, burning up over the Pacific Ocean.*

Timeline

Volume numbers (in **bold**) and page numbers are given at the end of each entry.

- **1,800,000 B.C.** The earliest **tools** yet discovered are made around this time in the Olduvai Gorge, Tanzania (**7**:5).
- **600,000–500,000 B.C.** Humans discover **fire** (**2**:53).
- **30,000 B.C.** The first **baked clay objects** are made (**7**:12).
- **12,000 B.C.** The first animals are **domesticated** (**6**:5).
- **10,000 B.C.** The first **clay pots** are made (**7**:12).
- **c. 8000 B.C.** In Europe **flint mines** are being worked (**7**:6).
- **5000 B.C.** Mesopotamians exploit wind power using **sails** (**6**:53).
- **4000 B.C.** **Glazes** are used to make pottery watertight (**7**:12).
- **3500 B.C.** **Bronze** is in use (**7**:22). **Wheels** are added to sleds to make the first **carts** (**6**:6).
- **3300 B.C.** The earliest known **writing system** is developed by the Sumerians in southern Mesopotamia (**3**:5).
- **3000 B.C.** In Egypt **glass** is used to make decorative beads (**7**:15). The **abacus** is invented (**5**:9). The **plow** is in use (**4**: 5, **10**:6).
- **2900 B.C.** Earliest **dam** is built on the Nile River in Egypt (**2**:35).
- **2700s B.C.** Early **stone buildings** date from this period (**2**:8).
- **2640 B.C.** **Silk** is discovered in China (**7**:45).
- **1900 B.C.** The Mesopotamians develop the **spoked wheel** (**6**:7).
- **1500 B.C.** The first **alphabet** is developed by Semitic peoples of the eastern Mediterranean (**3**:6). **Iron** smelting is first carried out on a large scale (**7**:22). The Egyptians make **body armor** by attaching small pieces of bronze to leather garments (**9**:7). The peoples of Egypt, Greece, and Phoenicia start to use glass to make **bottles** (**7**:15).
- **1300 B.C.** The Hyksos invent the **horse bit**, a piece of metal placed between the horse's teeth that gives the rider more control (**6**:8).
- **1000 B.C.** Babylonians and Egyptians develop **calendars** (**5**:16). Underground **water-storage facilities** are built in Jerusalem (**2**:33).
- **600 B.C.** Chinese develop the **blast furnace** to make iron (**7**:23).
- **430 B.C.** Greek physician Hippocrates develops a new medical technique called **clinical observation** (**8**:8).
- **250 B.C.** Eratosthenes draws a map of the world that includes lines of **longitude** and **latitude** (**5**:61).
- **100 B.C.** Syrians develop the technique of **glass blowing** (**7**:15).
- **46 B.C.** The **Julian calendar** is introduced (**5**:17).
- **A.D. 62** The earliest known **cookbook** is published (**2**:55).
- **105** In China **writing paper** is produced using mulberry bark (**3**:7).
- **600** Large **windmills** are built in Persia (**10**:11). Texts and pictures **printed** from carved wooden blocks first appear in China (**3**:8).
- **800** The Chinese produce **porcelain** (**7**:12).
- **868** The *Diamond Sutra*, the oldest known **printed book**, is produced in China (**3**:7).
- **1000** First known formula for **gunpowder** appears during this century in a Chinese manuscript (**9**:14).
- **1040** Reusable pieces of type are used for printing in China (**3**:8).
- **1100s** The first true **sailing warship**, the cog, is built (**9**:47).
- **1300** The **crown glass** process is used in Normandy, France (**7**:15).
- **1400** **Metal guns** are used for the first time (**9**:15).
- **1450** Goldsmith Johannes Gutenberg opens the first **printing press** in Mainz, Germany (**3**:11).
- **1490** A **bicycle** with two wheels, pedals, and a chain drive to the back wheel is designed by Leonardo da Vinci (**6**:18).
- **1494** Monk John Cor distills **whiskey** for his abbey (**4**:60).
- **1543** Andreas Vesalius makes many advances in the study of **anatomy** and publishes his findings in his famous book, *On the Fabric of the Human Body* (**8**:10).
- **1550** **Railroads** are invented in Europe for use in mines (**6**:25).
- **1568** Nikolaus Rensperger invents a universal **compass** (**5**:57).
- **1592** Galileo Galilei makes the first **thermometer** (**5**:30).
- **1608** Hans Lippershey invents the **telescope** (**5**:25).
- **1628** Physician and anatomist William Harvey discovers the true nature of the **circulation of blood** inside the body (**8**:10).
- **1640** Evangelista Torricelli builds the first **barometer** (**5**:41). The **bayonet** is invented (**9**:18).
- **1656** Christiaan Huygens designs the first **pendulum clock** (**5**:19).
- **1660** Robert Hooke builds a **compound microscope** (**5**:24).
- **1712** Thomas Newcomen develops a **steam engine** for removing water from mines (**7**:8, **10**:30).
- **1714** Daniel Gabriel Fahrenheit devises the **Fahrenheit temperature scale** (**5**:31, **8**:37).
- **1736** Joshua Ward invents a way of producing **sulfuric acid** (**7**:36).
- **1743** Astronomer Anders Celsius invents the **Celsius temperature scale** (**5**:31, **8**:36).
- **1775** John Outram invents the **streetcar** (**6**:30).
- **1779** The first **iron bridge** in the Western world is constructed at Coalbrookdale in England (**2**:19, **7**:24).
- **1780** Antoine Lavoisier and Pierre Simon Laplace invent the **calorimeter** (**5**:32).
- **1782** James Watts invents the double-acting **steam engine**, an improvement on his engines that produce rotary motion (**10**:33).
- **1783** Montgolfier brothers ascend in a **hot-air balloon** (**1**:6). Jacques-Alexandre Charles ascends in a **hydrogen balloon** (**1**:7).
- **1784** Henry Cort patents the **puddling process** for making wrought iron (**7**:24).
- **1787** Oliver Evans builds first **fully automatic** mill (**7**:56, **10**:15).
- **1794** Jacob Schweppe introduces the **fizzy drink** (**4**:63).
- **1795** Nicolas Appert develops a new system of **preserving food** by heating it in sealed containers (**4**:54).
- **1798 Lithography** is invented by Aloys Senefelder (**3**:9).
- **1800** Alessandro Volta invents the **electric battery** (**10**:46).
- **1801** Joseph-Marie Jacquard devises an **automatic loom** (**7**:48). Richard Trevithick builds a **steam-powered carriage** (**6**:37, **10**:33).
- **1804** Sir George Cayley constructs the first working model of a **fixed-wing aircraft** (**1**:17).
- **1812** The first practical **steam railroad** starts operating between Leeds and Middleton in England (**6**:26).
- **1816** René Laënnec makes the first **stethoscope** (**8**:37).
- **1820s** Joseph Henry and William Sturgeon produce the first working **electromagnets** (**10**:51).
- **1821** Charles Babbage constructs the **Difference Engine**, the ancestor of modern computers (**3**:53).
- **1824 Portland cement** is patented by Joseph Aspdin (**2**:12).
- **1826** Chemist Joseph Niépce produces the first **positive photographic image** (**3**:25). The **gas stove** is designed by James Sharp (**2**:53–54). Thomas Telford builds the first modern **suspension bridge** (**2**:46).
- **1829** George Stephenson unveils his **steam-powered locomotive**, *Rocket*, designed mainly by his son Robert (**6**:26).
- **1831** Scientists in the United States, Germany, and France discover the anesthetic properties of **chloroform** (**8**:17).
- **1835** Samuel Colt patents the Colt **revolver** (**9**:19).
- **1836** Francis Smith and John Ericsson come up with a successful design for a **screw propeller** (**6**:56, **9**:48).
- **1837** Samuel Morse develops an electric **telegraph** system that uses a code known as **Morse code** (**3**:15).
- **1839** Charles Goodyear finds that rubber can be made more stable by heating with sulfur, a process later named **vulcanization** (**7**:38).
- **1842** Surgeon Crawford Long carries out the first **pain-free operation** using ether as an anesthetic (**8**:16).
- **1844** John Gorrie builds a **refrigerator** (**4**:56).
- **1848** The **breech-loading rifle**, invented by Johann Dreyse, comes into service (**9**:21).
- **1849** The **block signaling system** for railroads is invented (**6**:29).
- **1850s** Oscar Levi Strauss invents modern **jeans** (**7**:50). Amelia Bloomer invents **bloomers**—pants for women (**7**:51).
- **1850s–1880s** Microbiologists Louis Pasteur and Robert Koch discover that **microorganisms** cause many **diseases** (**8**:17–18).
- **1854** Elisha G. Otis develops a safe **elevator** (**2**:22).
- **1855** Alexander Parkes creates a **plastic** called Parkesine (**7**:39).
- **1856** Henry Bessemer invents the **Bessemer converter** for making steel cheaply (**7**:24). Louis Pasteur develops his system of **pasteurization** (**4**:61). William Henry Perkin creates the first **synthetic dye** (**7**:34).
- **1860s** Pierre Lallement constructs the first bicycle with pedals. The machine is called a **velocipede** (**6**:19). Gregor Mendel studies plant breeding and lays the foundations of **genetics** (**4**:41).
- **1865** Joseph Lister performs the first **antiseptic operation** (**8**:22).
- **by 1867** Nikolaus August Otto makes the first working **internal-**

combustion engine that uses the **four-stroke cycle** (10:36). The first modern **cantilever bridge** is built (2:46).

- **1868** Christopher Sholes and Carlos Glidden patent their **typewriting machine** (3:7).
- **1869 Margarine** is invented by Hippolyte Megè-Mouriès (4:59).
- **1870s** The **assembly line**—a mechanized production technique—first appears in the meat-packing factories of Chicago (7:57).
- **1873** The first commercial **electric motors** are built by Zénobe T. Gramme (10:52).
- **1875** Chemist Alfred Nobel invents **dynamite** (9:25).
- **1876** Alexander Graham Bell and Elisha Gray independently file patents for **telephones** (3:16). Karl von Linde designs the first **domestic refrigerator** (2:56).
- **1877** Thomas Edison launches the **phonograph** (3:33).
- **1878** The first **telephone exchange** opens in Connecticut (3:16).
- **1878–1879** Thomas Edison develops the **light bulb** (10:27).
- **1879** Inventors Werner von Siemens and Johann Georg Malske unveil their **electric locomotive** (6:30).
- **1881** Louis Pasteur demonstrates a **vaccine** against anthrax (8:21).
- **1882** The first **hydroelectric power plant** is built at Appleton, Wisconsin (10:61). Coal-fueled **power plants** are built at Holborn Viaduct in London and at Pearl Street in New York to generate electricity (10:53).
- **1883–1885** The first **skyscraper** is constructed in Chicago (2:22).
- **1884** The **fountain pen** is invented by Lewis Waterman (3:6). Hiram Maxim develops the first fully automatic **machine gun** (9:26).
- **1885** Gottlieb Daimler builds the first modern **motorcycle** (6:39).
- **1886** John Pemberton invents **Coca-Cola** (4:63). Thomas Edison invents the **carbon microphone** (3:17).
- **1887** Heinrich Rudolf Hertz finds that light can be used to generate electricity, a phenomenon called the **photoelectric effect** (10:63).
- **1888** Showman Louis Le Prince shoots what is probably the first **movie**, a view of Leeds Bridge in England (3:29). The **pneumatic tire** is invented by John Boyd Dunlop (6:21). Inventor George Eastman brings out the first mass-produced **camera**, the Kodak no. 1 (3:27).
- **c. 1890** The **zipper** is invented by Whitcomb Judson (7:52).
- **1891** Count Hilaire de Chardonnet's **rayon** becomes the first commercially produced artificial fiber (7:51).
- **1892** François Hennebique patents **reinforced concrete** (2:25). Rudolf Diesel patents the **diesel engine** (10:36).
- **1895** Physicist Wilhelm Röntgen discovers **X-rays** (8:38). Engineer Guglielmo Marconi uses **radio** to transmit telegraph signals over a distance without wires (3:18).
- **1897** Karl Braun invents the **cathode-ray tube** (3:43).
- **1901** Hubert Cecil Booth designs the first **vacuum cleaner** (2:62).
- **1902 Trinitrotoluene (TNT)** is invented in Germany (9:25).
- **1903** Bicyclemakers Wilbur and Orville Wright achieve the world's first **controlled flight in a powered airplane** (1:19). Schoolteacher Konstantin Tsiolkovsky suggests using **liquid-fuel rockets** to reach space (1:42).
- **1905** Albert Einstein sees the potential of **nuclear power** (10:57).
- **1907** Alva J. Fisher invents an **electric washing machine** (2:61, 10:55). Boris Rosing builds the first working **television system** (3:43).
- **1908** Henry Ford starts **mass producing** the Model T, bringing **automobiles** within the reach of ordinary people (6:43, 7:58).
- **1909** Fritz Haber invents a process for making **ammonia** (7:39).
- **1910** Salvarsan, the first **synthetic chemical cure**—discovered by Paul Ehrlich and Sahachiro Hata—begins to be sold (8:26).
- **1912** The first **diesel locomotive** is built (6:31).
- **1914** A team at Johns Hopkins Hospital in Baltimore make the first **kidney dialysis machine** (8:47).
- **1915** C. Sullivan and W. C. Taylor create a type of **heat-resistant glass** called Pyrex (7:17). Engineers Ernest Swinton and Walter Tritton build the first practical **tank** (9:30).
- **1920s** The **factory ship** is introduced to the fishing industry (4:23).
- **1924** Clarence Birdseye's company is making **frozen food** (4:56).
- **1926** Engineer John Logie Baird gives his first public demonstration of **television** (3:42). The first **liquid-fuel rocket** to fly is designed by Robert Hutchings Goddard (1:43).
- **1928** Alexander Fleming discovers penicillin, an **antibiotic** (8:28).

- **1930** Frank Whittle patents the **jet engine** (1:29). Lewis Richardson develops **echo sounding** in ships (5:46). John H. Gibbon invents a **heart-lung machine** (8:48).
- **1933** Boeing designs the first modern **airliner** (1:25).
- **1935** Wallace Hume Carothers invents **nylon** (7:42, 7:53).
- **1936** The BBC starts regular **live television broadcasts** from Alexandra Palace in London (3:44).
- **1938** Georg and Lazlo Biro produce their **ballpoint** pen (3:6). Chester Carlson invents **xerography** (3:10).
- **1939** The first **jet-powered airplane**, the Heinkel He-178, is flown (1:30). The first practical **helicopter** is flown in the United States by engineer Igor Sikorsky (1:38).
- **1942** Enrico Fermi sets up the first self-sustaining **nuclear chain reaction** (10:58).
- **1944** Massey Harris launches a **combine harvester** (4:28).
- **1945** The first **atomic bombs** are dropped by the United States on the Japanese cities of Hiroshima and Nagasaki (9:33).
- **1947** Edwin Land brings out the first **instant camera**, the Polaroid Land camera (3:28).
- **1948** Scientists at Bell Laboratories invent a type of miniature electronic switch called a **transistor** (3:18, 3:58, 10:47).
- **1952** Alistair Pilkington develops the **float glass method** for producing sheet glass (7:16). The first **hydrogen bomb** is tested (9:33). Physician Robert Lee Wild produces the first **ultrasound images** of structures inside the body (8:38).
- **1953** The structure of **deoxyribonucleic acid (DNA)** is determined by James Watson and Francis Crick (4:44). The United States develops a **surface-to-air missile (SAM)** by this time (9:39).
- **1956 Transistorized computers** are developed at Bell Laboratories in the United States (3:58).
- **1957** The first **artificial satellite**, *Sputnik I*, is launched (1:44–45).
- **1958** Rune Elmqvist makes the first **heart pacemaker** (0.50).
- **1959** Scientists working at Texas Instruments develop the **integrated circuit** (3:59, 10:47). The **hovercraft** is invented by Christopher Cockerel (6:63).
- **1960** The first **all-transistor TV** is produced by Sony (3:46).
- **1961** The USSR (now the CIS) launches the first **passenger-carrying spacecraft** with astronaut Yuri Gagarin on board (1:46).
- **1962** *Telstar*, the first working **communications satellite**, is launched (3:21). The first **industrial robot** is developed by George C. Devol and Joseph Engelberger and sold to General Motors (7:62).
- **1963** The first **biological laundry powder** is launched (4:42).
- **1967** Surgeon Christiaan Barnard carries out the first human-to-human **heart transplant** (8:50). A **magnetic levitation (maglev) train** is constructed and tested in Colorado (6:34).
- **1969** NASA launches *Apollo 11*, the first successful **crewed mission to the Moon** (1:47–49).
- **1970** Japanese companies Sony, JVC, and Matsushita bring out the fist **videocassette recorders** (3:49).
- **1971** Engineer Ted Hoff invents the **microprocessor** (3:59, 10:47).
- **1972** Janet Mertz and Ron Davis use cut-and-paste techniques to make the first **recombinant DNA** engineered by humans (4:46).
- **1977** Raymond Damadian takes the first pictures inside the human body using **magnetic resonance imaging (MRI)** (8:40).
- **1979 Mobile phones** are introduced by AT&T (3:21). Philips and Sony develop the **compact disc** (3:38).
- **1981** NASA launches the first **space shuttle**, *Columbia* (1:53).
- **1983** The **Stealth fighter**, or F-117A, is built by Lockheed (1:35).
- **1987 Digital audio tape (DAT)** recorders go on sale (3:38).
- **1989** The **World-Wide Web (WWW)** is invented by computer scientist Tim Berners-Lee (3:63).
- **1994** The first **genetically engineered food crop**, the Flavr Savr tomato, reaches the public (4:50).
- **1995** *Toy Story* becomes the first **full-length film to be entirely computer generated** (3:31).
- **1997** Dr. Ian Wilmut leads a team of scientists who produce the first **clone** of an adult mammal, Dolly the sheep (8:63).
- **1998** Construction of **International Space Station** begins (1:50).
- **2000s** The **Human Genome Project** aims to have determined the chemical sequence of every human gene during this century (8:62).

Glossary

aerodynamic Able to move through a fluid (such as air) efficiently. Aerodynamic objects are shaped to reduce **drag**.

airfoil A shaped surface, typically curved on one side and straight on the other, that produces both lift and **drag** when moved through the air. Airplane wings and propellers are examples of airfoils.

algebra A system that uses symbols (often letters) to represent numbers and operations, allowing mathematical relationships to be expressed in general terms.

alternating current A type of electric current in which the direction of flow changes at regular intervals.

American Revolution (1775–1783) A war fought between Great Britain and the 13 British colonies in North America. The colonies were victorious and went on to form the United States.

amplify To make an electronic signal larger or stronger. **Amplifiers** are devices that amplify signals and are widely used in electronic equipment such as televisions, computers, and radios. **amplification**—noun

amplitude The maximum range of an oscillation (fluctuating quantity), such as the height of a wave. For example, the amplitude of sound waves reaching the ear determines the perceived loudness of a sound.

analog A system of measurement in which one continuously varying quantity is measured in terms of a corresponding continuously varying quantity. A mercury thermometer is an example of an analog device since it measures temperature variations in terms of the height of mercury in a glass tube.

anatomy The arrangement of body parts, such as tissues and organs, inside the body of a living organism, and the science of identifying and describing these parts.

ancient Greece A civilization that existed on the mainland and islands of modern-day Greece and Turkey between 2000 and 300 B.C.

anode A positively charged **electrode**, which attracts negatively charged particles.

antibody A protein produced by the body's immune system in response to foreign substances, or **antigens**. Many different types of antibodies are made, each able to bind to a particular antigen. Once antibodies have bound, the antigen can be engulfed and destroyed by **white blood cells**.

antigen A protein, toxin, or other large **molecule** to which the body reacts by producing **antibodies**.

artillery Large guns designed to fire heavy projectiles such as explosive shells, or the study and use of such guns. Artillery pieces can be mobile, stationary, or mounted on ships or airplanes and are often capable of destroying targets many miles away.

atom The smallest unit of a chemical element that can take part in a chemical reaction. Atoms are composed of a central nucleus made up of **protons** and **neutrons** surrounded by shells of **electrons**.

bacteria Single-celled **microorganisms** that are present almost everywhere on Earth. Bacteria can only be seen through a microscope, and most are spherical, rodlike, or spiral in shape. Some bacteria are beneficial to people, helping us digest our food and playing an important part in the preparation of some foods. Others, however, are responsible for serious diseases such as cholera.

binary code A numbering system that uses only the digits one and zero to represent all numbers. Binary code is widely used in electronics and computers since it can be processed quickly using simple two-way switching circuits such as logic gates.

buttress A structure usually made from brick or stone that is built against a wall to support and reinforce it.

catalyst A substance that increases the rate of a chemical reaction without undergoing a permanent chemical change itself.

cathode A negatively charged **electrode**, which attracts positively charged particles.

cathode-ray tube Glass tube with all the air removed (leaving a **vacuum**) that encloses a beam of high-speed, high-energy **electrons**.

centrifugal force A force that seems to pull a thing outward as it rotates around an axis.

Civil War (1861–1865) Also called the American Civil War, this conflict was fought between the Northern (Union) and the Southern (Confederate) states. One of the major disputes was over slavery, which the South supported and the North wanted to stop. The North won, and slavery was made illegal throughout the United States.

classical From either **ancient Greece** or ancient **Rome**.

Cold War An extended period of hostility that ran from 1945 to the start of the 1990s. A group of communist states, led by the Soviet Union (now CIS), formed one side, and they were opposed by a group of noncommunist nations, dominated by the United States. Although conflict never broke out, there was a constant threat of war, and both sides produced vast numbers of nuclear and conventional weapons in case of attack. The Cold War ended with the collapse of many communist governments in the late 1980s and early 1990s.

compression The act of being made more compact by the application of pressure.

conductor Any material that can easily transmit heat or electricity.

decimal A numbering system based on the number ten and using digits from zero to nine.

digital A measuring system, such as **binary code**, that can store information as a number of discrete values. A binary computer is an example of a digital device since it uses only two values—one and zero—to store data.

direct current An electric current that flows in a constant direction.

domesticated Animals or plants that have been adapted by selective breeding over many generations to live in close association with humans and perform some useful function are domesticated. **domestication**—noun

drag A force that opposes the motion of an object through a fluid such as air or water.

electrode A **conductor** that carries electric current into or away from the other conducting substances in an electric circuit. Electrodes are often used to make electrical contacts with nonmetals.

electromagnet An iron core surrounded by a coil of wire that temporarily generates a **magnetic** field when an electric current flows through the wire. This phenomenon is called **electromagnetism**.

electromagnetic radiation/waves Waves made up of electric and **magnetic** fields. Electromagnetic waves travel at the speed of light, but the characteristics of a particular wave are determined by its **frequency**. **Radio waves** and **microwaves** are low-frequency electromagnetic waves, while **infrared**, visible light, **ultraviolet**, and **X-rays** are produced as the frequency increases.

electrons Negatively charged particles that are basic constituents of **atoms**. Electrons are involved with many phenomena such as the flow of electric currents through circuits.

enzyme A biological **catalyst**—a protein molecule produced by an organism to control or perform a particular chemical reaction.

fermentation The breakdown of sugars to release energy. In certain conditions some types of yeast and other **microorganisms** release carbon dioxide and alcohol as waste products from fermentation, an effect that is used in breadmaking and brewing.

fiber-optic cables Cables made up of many transparent fibers (often made from glass or plastic) through which light from a laser is transmitted. Fiber-optic cables can carry large amounts of information in the form of light pulses at high speeds over great distances.

frequency The number of oscillations (movements between peak and trough) of a wave over a fixed amount of time. Usually this time period is one second, and the frequency is measured in units called hertz.

friction A force that resists the relative motion of two bodies in contact.

gamma rays High-energy electromagnetic radiation with a wavelength shorter than 0.01 nm (billionths of a meter). Gamma rays are emitted after nuclear reactions.

genetic Relating to, or the study of, the inherited features of an organism.

global warming The gradual warming of Earth due to heat from the Sun being trapped by certain gases building up in the atmosphere. The most important of these gases is carbon dioxide, and its levels are rising largely because of human activities such as the burning of coal and oil.

gravity A natural force that attracts two masses toward one another. Among its many effects, gravity draws objects toward Earth's surface and keeps the planets in **orbit** around the Sun.

gyroscope A wheel mounted in a set of rings so that its axis of rotation is free to turn in any direction. When the wheel is spun rapidly, the direction of the axis of rotation will remain constant no matter which way the gyroscope is turned.

hertz See **frequency**

Holocaust The systematic murder of six million Jews and thousands of gypsies, Poles, and Russians by the **Nazis** during **World War II**. Victims of the Holocaust were executed, starved, or worked to death, often at specially constructed concentration camps.

hormone A substance released by certain glands into the bloodstream that travels to and acts on remote sites in the body. Both natural hormones and artificial hormones are often used in medicine.

hydraulic Operated by the movement and force of a liquid. Most hydraulic systems consist of a series of liquid-filled tubes through which a force can be transmitted.

immunize To protect from the effects of a harmful substance or disease. There are two forms of immunization—passive, which involves injecting the patient with **antibodies**, and active, in which a form of harmful substance that has been made harmless in some way is injected, allowing the patient to generate antibodies toward it.

Industrial Revolution A great change in social and economic organization brought about by the replacement of hand tools by machines and power tools, and the development of large-scale industrial production methods. The Industrial Revolution started in England around 1760 and spread to the rest of Europe and the United States.

infrared a type of **electromagnetic radiation** covering wavelengths between 0.75 and 10,000 µm (millionths of a meter). Infrared rays have a strong heating effect and are emitted by hot objects.

Internet An international network connecting millions of computers. A piece of software called a browser on one computer is used to access information stored on another computer, called a server, which may be thousands of miles away.

ion A charged particle created when a neutral **atom** or **molecule** loses or gains an **electron**.

irradiation Exposure to a form of **electromagnetic radiation**, such as **X-rays**, **gamma rays**, or **ultraviolet** light.

Korean War (1950–1953) A war in which a U.S.-dominated United Nations coalition came to the aid of South Korea during an invasion by North Korea, which was aided by the Soviet Union (now CIS) and communist China. The war was ultimately indecisive, but was one of the key military confrontations of the **Cold War**.

liquid crystal A fluid that exists in a state between being liquid and forming solid crystals. Some types of liquid crystal change from being see-through to scattering light when an electric current is applied and are used in electronic displays.

liquid-fuel rocket A type of rocket engine in which the propellants are liquids. The fuel, often liquid hydrogen, and oxidant, liquid oxygen, are stored in separate tanks and pumped into the combustion chamber, where they are ignited.

magnetism All phenomena associated with magnets and **magnetic fields**. Magnetic fields are regions around magnets in which a force acts on any magnet or electric charge present.

magnets Any material capable of generating a magnetic field. See also **magnetism**.

matter Any material substance that occupies space and is observable in some way. **Atoms** are the basic building blocks of matter, and all matter is subject to the force of **gravity**.

medieval Of or from the **Middle Ages**.

metallurgist A person who studies the science of metals, particularly the extraction and processing of metal ores.

meteor The visible track seen when a meteoroid (a small, solid body from outer space) enters Earth's atmosphere.

meteorite Any fragment of a meteoroid that survives its passage through the atmosphere and falls to the surface of a planet or moon as a lump of rock or metal. See also **meteor**.

meteorologist A scientist who studies the atmosphere and atmospheric phenomena and tries to predict the weather.

microorganism An organism too small to be seen with the eye alone. For example, **bacteria**, **viruses**, and yeasts.

microwave An **electromagnetic wave** with a wavelength between 1 mm and 300 mm. Microwaves are used for radar, communications, and heating foods.

Middle Ages A period of European history that ran from around A.D. 500 to c. 1450.

Nazi The name of a political party or its members that, led by Adolf Hitler, ruled Germany from 1933 to 1945. The Nazis suppressed all opposition and built up Germany's military strength. They were the main protagonists of **World War II**. See also **Holocaust**.

neutron A component of an **atom**'s nucleus. Neutrons do not have an electric charge.

nomads People who, instead of having a fixed home, travel from place to place in search of fresh pastures and water for themselves and their animals.

optics The study of light and vision.

orbit The path, shaped like a circle or an ellipse, that an object in space takes around another object.

photo diode An electronic component that is used to detect or measure the brightness of light. Photo diodes are usually made from semiconductors, such as silicon.

photoelectric cell A tube enclosing a **cathode** made from a material that emits **electrons** when it is exposed to light and an **anode** that collects these electrons. In a circuit containing a photoelectric cell current flows when the cell is exposed to light.

piezoelectric Able to produce electricity when subjected to a mechanical force. The piezoelectric effect is found in certain types of crystal. **piezoelectricity**—noun

positrons Positively charged subatomic particles with the same mass as **electrons**. Positrons are very unstable and rapidly react with any **matter** to form **gamma rays**.

protons Basic components of **atoms**. Protons are located in the nucleus and have a positive electric charge.

radiation The act of giving off **radioactive** particles, heat, or **electromagnetic waves**.

radio waves A type of **electromagnetic wave** with a wavelength between 1 mm and 1,000 m. Radio waves are used by people to transmit information over long distances.

radioactivity The disintegration of atomic nuclei accompanied by the giving off of particles or **electromagnetic waves**. **radioactive**—adjective

recombinant DNA DNA molecules that have been created by artificially pasting together DNA fragments that are often taken from different species.

refrigerant A fluid used to transfer heat from one space to another. The refrigerant is usually pumped around a closed circuit, absorbing heat from the cooler space and releasing it in the warmer space.

remote sensing The measuring of **electromagnetic radiation**. Remote-sensing devices, often mounted on aircraft or **satellites**, usually detect electromagnetic waves reflected from the surface of Earth.

Rome The ancient civilization that began in the Italian city of Rome around 700 B.C. and had established a vast empire around the Mediterranean by 200 A.D. The Romans are noted for being the first to bring law and order to Europe and for their great works of engineering. **Roman**—noun, adjective

satellite A natural or artificial object in **orbit** around a star, planet, or other body.

solar panels Collections of solar cells used to convert sunlight into electrical energy. They are often used to provide electricity on spacecraft.

Solar System The Sun together with the nine planets (including Earth) and other bodies (such as asteroids) that **orbit** it.

solid-fuel rocket A rocket in which the fuel and oxidant are both solids. Typically, solid-fuel rockets are simple metal tubes packed with a mixture of the propellants that are ignited to provide thrust. Solid-fuel rockets are less complex but more difficult to control than **liquid-fuel rockets**.

sound barrier The abrupt increase in **drag** that occurs when an object approaches the speed of sound.

Soviet Of or from the USSR, a communist state that existed from 1923 to 1989 and included present-day Russia.

superconductor A material that can conduct electricity without any resistance. The few materials that exhibit superconductivity only do so at extremely low temperatures.

supersonic Traveling or capable of traveling faster than the speed of sound.

telecommunications Sending messages over a distance, usually involving electrical signals or **electromagnetic waves**.

tension A force that opposes attempts to deform an object. A piece of elastic being stretched is under tension.

thermodynamic Of or related to the branch of physics that deals with the conversion of heat to and from other forms of energy. It applies particularly to devices that use heat to generate motive power.

torque An external force that causes an object to turn.

torsion The twisting of an object by exerting a force to turn one part in one direction while another part is held stationary or turned in the opposite direction.

transmission An assembly of parts in a motor vehicle, including gear wheels and shafts, that transmits power from the engine to the axles and wheels.

turbine A machine made up of a set of blades mounted on a central shaft. A moving fluid, such as steam or air, makes the assembly rotate. Turbines are often used to drive generators.

ultraviolet A part of the **electromagnetic** spectrum covering wavelengths between 10 and 380 nm (billionths of a meter). Ultraviolet light is invisible to the human eye.

vaccine A medication, usually containing disease-causing **microorganisms** that have been killed or weakened, which is administered to **immunize** or increase immunity to a particular illness.

vacuum A space that contains no **matter**.

Vietnam War (1957–1975) A conflict between communist North Vietnam (supported by the Soviet Union and China) and noncommunist South Vietnam (supported by the United States). U.S. soldiers were actively involved in the war from around 1965 to 1973, after which hostile public opinion in the United States forced their withdrawal. North Vietnam conquered the South in 1975.

virus A tiny, disease-causing particle that consists of protein combined with **genetic** material (DNA or RNA—ribonucleic acid). Viruses are only capable of replicating inside living cells, and for this reason many scientists do not consider them to be living organisms; other scientists, however, consider them to be a type of **microorganism**.

white blood cells Components of the blood responsible for destroying foreign substances and organisms. Different types of white blood cells perform different functions in this process, which is called the immune response. Some generate **antibodies**, while others engulf and digest foreign particles.

World War I (1914–1918) A war fought mainly in Europe between the Central Powers—Germany, the Austro-Hungarian Empire (present-day Austria and Hungary), and the Ottoman Empire (now Turkey)—and the Allies: France, the British Empire, Russia, and the United States. The Allies eventually won the conflict, but millions of soldiers on both sides lost their lives.

World War II (1939–1945) The most destructive conflict in history, fought mainly in Europe, East Asia, and North Africa. The Axis powers (Germany, Austria, Japan, and Italy) were opposed by the Allies (Britain, the United States, France, and the USSR). Germany surrendered in April 1945, but Japan fought on until August, when atomic weapons dropped by U.S. aircraft destroyed the Japanese cities of Hiroshima and Nagasaki.

X-ray A type of **electromagnetic radiation** with a wavelength between 0.001 and 10 nm. X-rays are able to travel through soft tissues and can be used to study the internal structures of the body.

Further Reading List

Brauch, H. G. (ed.), *Controlling the Development and Spread of Military Technology: Lessons from the Past and Challenges for the 1990s*, Paul & Co. Publishing Consortium: Concorde, MA (1992).

Brophy, A., *John Ericsson and the Inventions of War*, Silver Burdett Press: Parsippany, NJ (1991).

Buderi, R., *The Invention that Changed the World: How a Small Group of Radar Pioneers Won the Second World War and Launched a Technical Revolution*, Touchstone Books: New York, NY (1998).

Dale, R. & Weaver, R., *Discoveries and Inventions: Home Entertainment*, Oxford University Press: Oxford, UK (1994).

Day, T. and The Diagram Group, *Historical Inventions On File*, Facts On File: New York, NY (1994).

Dummer, G. W. A., *Electronic Inventions and Discoveries: Electronics from Its Earliest Beginnings to the Present Day*, IOP Publications/Institute of Physics: Bristol, UK (1997).

Eisen, J. (ed.), *Suppressed Inventions and other Discoveries: Revealing the World's Greatest Secrets of Science and Medicine*, Avery Publishing Group: Wayne, NJ (1999).

Fenichell, F., *Plastic: the Making of a Synthetic Century*, Harperbusiness: New York, NY (1997).

Fisher, D. E., *Tube: the Invention of Television*, Harcourt Brace: Orlando, FL (1997).

Gourley, C., *Wheels of Time: a Biography of Henry Ford*, Millbrook Press: Brookfield, CT (1997).

Grady, S. M., *Ships: Crossing the World's Oceans*, Lucent Books: San Diego, CT (1992).

Hoare, S., *20th-Century Inventions: Digital Revolution*, Raintree/Steck Vaughn: Austin, TX (1998).

Hooper, T., *Surgery*, Raintree/Steck Vaughn: Austin, TX (1993).

Jeffrey, L. S., *American Inventions of the Twentieth Century*, Enslow Publishers Inc.: Springfield, NJ (1996).

Karwatka, D., *Technology's Past: America's Industrial Revolution and the People who Delivered the Goods*, Prakken Publications: Ann Arbor, MI (1996).

Lafontaine, B., *Great Inventors and Inventions*, Dover Publications: Mineola, NY (1998).

Marawetz, H., *Polymers: the Origins and Growth of Science*, Dover Publications: Mineola, NY (1995).

Newhouse, E. L. (ed.), *Inventors and Discoveries: Changing Our World*, National Geographic Society: Washington, DC (1994).

Parker, S., *20th-Century Inventions: Satellites*, Raintree/Steck Vaughn: Austin, TX (1997).

Platt, R., *Inventions: a Visual History*, Dorling Kindersley: London, UK (1994).

Pollard, M., *The Clock and How It Changed the World*, Facts On File: New York, NY (1995).

Pollard, M., *The Lightbulb and How It Changed the World*, Facts On File: New York, NY (1995).

Pratt, P. B., *Maps: Plotting Places on the Globe*, Lucent Books: San Diego, CA (1995).

Steffens, B., *Printing Press: Ideas into Type*, Lucent Books: San Diego, CA (1990).

Stengers, I., *Power and Invention: Situating Science*, University of Minnesota Press: Minneapolis, MN (1997).

Turvey, P., *Inventions: Inventors and Ingenious Ideas*, Franklin Watts Inc.: Danbury, CT (1994).

Usher, A. P., *A History of Mechanical Inventions*, Dover Publications: Mineola, NY (1988).

Set Index

Volume numbers are in **bold**. Page numbers in *italics* refer to picture captions, diagrams, or a caption and text reference on the same page.

A

abacus **3**:51, **5**:9
ABS system **6**:46–48
aerodynamics **1**:*16*, **6**:23
agriculture
  and biotechnology **4**:48–50
  modern **4**:27–37
  traditional **4**:5–17
AIDS **8**:28, **8**:33, **8**:58
air conditioning **2**:28
aircraft **1**:5–39, **4**:33, **5**:46, **5**:60, **10**:39
aircraft carriers **1**:34, **9**:51
airfoils **1**:15, **1**:*16*, **1**:22, **6**:55, **10**:38
airliners **1**:10, **1**:25, **1**:32, **1**:37
airplanes **1**:11, **1**:*15–25*, **4**:33
  autopilots **5**:59, **7**:62
  jet **1**:27–39, **10**:39
airships **1**:8–12, **1**:15
alchemy **7**:35
alkalis **7**:36–37
alloys **7**:22, **7**:26
alphabets, early **3**:6
aluminum **7**:11, **7**:27–28, **7**:29
ammonia **7**:37, **7**:39
amniocentesis **8**:*61*
Ampère, André-Marie **10**:51
analog data **3**:22, **3**:36, **3**:37
anatomy **8**:9–10
anemometers **5**:*38*
anesthetics **8**:15, **8**:16–17, **8**:32, **8**:45
  for childbirth **8**:56, **8**:57
animals
  and growing crops **4**:6–8, **4**:27
  livestock **4**:16–17, **4**:34–35, **4**:51
  for transport **4**:17, **6**:5–15
antibiotics **4**:25, **4**:36, **4**:37, **8**:28–30, **8**:32
antibodies **8**:20, **8**:46
Antikythera mechanism **10**:8
antiseptics/antisepsis **8**:22, **8**:45, **8**:56
Apollo Moon program **1**:47–49, **1**:50, **3**:46
Appert, Nicolas **4**:54, **4**:55
aquaculture **4**:*19*, **4**:24–25
aqueducts **2**:14, **2**:33, **8**:7
arches **2**:*13*, **2**:14, **2**:16
  for bridges **2**:42, **2**:43–45
  for dams **2**:36, **2**:37
Archimedes **6**:56, **10**:6, **10**:17
Aristotle **10**:8
armor **7**:22, **9**:7, **9**:10, **9**:14
Armstrong, Edwin H. **3**:18–22
Armstrong, William **2**:15, **9**:23
artificial insemination **4**:35–37
artillery **9**:11–13, **9**:23–24
aseptic surgery **8**:22, **8**:45
aspirin **8**:7, **8**:25, **8**:30, **8**:32
assembly lines **6**:42, **6**:43, **7**:57, **7**:58, **7**:60, **10**:29
astronomy **5**:6, **5**:25
  satellites (artificial) **1**:57–58
  and time measurement **5**:13–18
  See also Moon; Solar System; Sun
atoms **5**:34, **9**:33, **10**:59
automatic pilots **5**:59, **7**:62
automation **7**:27, **7**:55–63, **10**:*15*
automobiles **6**:39, **6**:40, **6**:41–49, **7**:16, **7**:27, **7**:31, **7**:58, **10**:36

B

Babbage, Charles **3**:53–54, **10**:10
Baird, John Logie **3**:41, **3**:42–43
bakelite **7**:40
baking **4**:40, **4**:53
ballistas **9**:11, **9**:12
ballooning **1**:5–8, **1**:12, **1**:13, **1**:15
  weather balloons **5**:43, **5**:47
barometers **5**:41, **5**:42
baths/bathrooms **2**:56–60, **8**:7
batteries **7**:27, **8**:50, **10**:44
  10:46–48, **10**:52
battering rams **9**:8–10
battleaxes **9**:6
Baudot code **3**:15
bayonets **9**:*18*, **9**:19
bazookas **9**:31
Becquerel, Antoine-Henri **5**:27, **5**:34, **10**:57
beer **4**:59–60
Bell, Alexander Graham **3**:16–17, **3**:33, **6**:62
Benz, Karl **6**:39, **6**:40, **6**:41
Berliner, Emile **3**:33–34
Bernoulli, Daniel **1**:16
Bessemer converter **2**:19, **7**:25, **7**:26
bicycles **6**:*18–23*, **6**:39, **7**:18
binary code **3**:21, **3**:*22*, **3**:37, **3**:38, **3**:39, **3**:56, **3**:57, **5**:8
biological warfare **9**:43
biometry **9**:63
biosensors **5**:52
biotechnology **4**:34, **4**:39–51, **5**:52–53
Birdseye, Clarence **4**:56–57
Biro, Georg and Lazlo **3**:6
bits (for horses) **6**:8–9
blacksmiths **7**:29
blast furnaces **7**:23, **7**:29
Blériot, Louis **1**:21
blood
  circulation **8**:10, **8**:11, **8**:45
  transfusions and blood groups **8**:45–46
boats *See* ships and boats
bombs
  atomic **9**:33, **9**:34, **10**:58
  disposal **7**:63
  hydrogen **9**:33–34, **10**:59
books and printing **3**:5, **3**:7–11
Boolean algebra **3**:56, **3**:57
bow and arrow **4**:19–20, **4**:*21*, **9**:6–8, **9**:10, **9**:14
Boyle, Robert **1**:5, **10**:30
brakes **6**:38, **6**:40, **6**:42, **6**:43–48
Bramah, Joseph **9**:58, **9**:59
brass **7**:22, **7**:27
Braun, Karl **3**:43
Braun, Wernher von **1**:29, **1**:42–44, **1**:57
breeding, selective **4**:19, **4**:35, **4**:39, **4**:40, **4**:48, **4**:49
breeds, animal **4**:14, **4**:35
bricks **2**:10–12
bridges **2**:41–47, **7**:24
bridles **6**:9, **6**:10
bronze **7**:22
Brunel, I. K. **6**:26, **6**:57–58
Brunel, Marc **2**:48, **2**:49–50

building materials **2**:6–12, **7**:19, **7**:33
  See also concrete; iron; steel
buildings
  modern **2**:19–31
  traditional **2**:5–17
  See also bridges; dams; tunnels
buoyancy **6**:56
buttons **7**:52

C

cable cars **6**:*31*
calculators **3**:51–55
calendars **5**:16–18
calorimeters **5**:32
cameras **3**:25, **3**:26, **3**:27–29
  television **3**:47
cams **10**:13, **10**:*14*, **10**:31
cancer, treatment **8**:32, **8**:33, **8**:51, **8**:62
canning **4**:54–55, **7**:27
cannons **7**:23, **9**:5, **9**:15–16, **9**:17
canoes **6**:51–52, **6**:63
Carlson, Chester **3**:10
Carothers, Wallace H. **7**:42
carpets **7**:51
carts **6**:6, **6**:12, **6**:*13*
Carson, Chester **3**:10
cash registers **10**:16
casting, metals **7**:30
catalysts **7**:39, **10**:23
catapults **9**:11, **9**:45
cathode-ray tubes **3**:43, **3**:48
cavalry **9**:13–14
Cayley, George **1**:*16*, **1**:17
celluloid **7**:40
Celsius, Anders **5**:31, **8**:36
ceramics **7**:12–14
CFCs **2**:57, **7**:40, **10**:*20*
chairlifts **6**:18
Chanute, Octave **1**:17
chariots **6**:8–9, **9**:*12*, **9**:13
Charles, Jacques-Alexandre **1**:7–8, **1**:*12*
chemicals **7**:33–43
chemical warfare **9**:27–28, **9**:43
chimneys **2**:17
chipboard **7**:11–12
chlorine **7**:37
chloroform **8**:16, **8**:17, **8**:*56*
cholera **2**:37, **8**:18
chromatography **5**:33
chromosomes **8**:*61*
chronometers **5**:20, **5**:59, **10**:9–10
chronophotography **3**:29
cinematography **3**:27, **3**:29–*31*
cities, planning **2**:29–31
city gas **10**:21, **10**:24, **10**:26
Clarke, Arthur C. **1**:46
clippers **6**:54
clocks **5**:18–21, **5**:25, **10**:8–9
  See also chronometers
cloning **8**:63
clothing **7**:45, **7**:50–53
coaches **6**:5, **6**:13–14, **6**:*15*
coal **7**:7, **7**:9, **10**:21, **10**:23
Coca-Cola **4**:63
Cockerell, Christopher **6**:61, **6**:63
coffee **4**:62–63
collar harnesses **4**:7, **6**:12, **6**:*13*
Colt, Samuel **9**:19
combine harvesters **4**:12, **4**:27, **4**:28
combustion **10**:*19*
compact discs (CDs) **3**:38
  CD-ROMs **3**:62
compasses (magnetic) **5**:25, **5**:55, **5**:56, **5**:*57*, **5**:60, **5**:62
composite materials **1**:53, **7**:18–19
computers **3**:20, **3**:29, **3**:*31*, **3**:53, **3**:56–63, **7**:60

in automobiles **6**:48, **6**:49
  and printing **3**:11
  and warfare **9**:34–36
  See also binary code
concrete **2**:12, **2**:25, **2**:36, **7**:9
Congreve, William **1**:42, **9**:49
containers, freight **6**:32
contraception **8**:31, **8**:51, **8**:57, **8**:58–59
Cooke, William **3**:14–15
cooking **2**:53–56
copper **7**:27, **7**:33, **9**:50
Corbusier, Le **2**:26–27, **2**:30
Cort, Henry **7**:29
cotton **4**:9, **7**:45–46
cotton gins **7**:46, **7**:56
Coulomb, Charles-Augustin de **5**:50
counting systems **5**:7–9
cranes **2**:*15*, **10**:17
cranks **6**:19, **10**:*10*, **10**:13–14, **10**:31
crankshafts **10**:14, **10**:35
Crick, Francis **4**:44
crop rotation **4**:12–16, **4**:17
cryosurgery **8**:52–53
CT (computed tomography) **8**:39, **8**:40, **8**:42
Cugnot, Nicholas **6**:37
Curie, Marie **10**:57, **10**:*58*
Curie, Pierre **5**:45, **10**:57
Curtiss, Glenn **1**:21, **1**:24
cylinders (engine) **6**:38, **6**:47, **7**:27

D

daguerrotypes **3**:25, **3**:26
Daimler, Gottlieb **6**:39–41, **10**:36
dams **2**:35–38, **4**:10
Darwin, Charles **4**:40
Davy, Humphry **7**:7, **7**:39, **8**:15, **8**:16, **10**:26, **10**:48, **10**:49
DDT **4**:32–33
Deere, John **4**:7, **4**:28
De Forest, Lee **3**:18, **3**:22, **3**:57
deoxyribonucleic acid (DNA) **4**:44–45, **8**:*61*
  fingerprinting **5**:53, **8**:*62*
  recombinant **4**:46
derailleur gears **6**:21, **10**:7
desktop publishing **3**:11, **3**:61
dialysis machines **8**:47–48
diesel locomotives **6**:31
digital technology **3**:22, **3**:28–29, **3**:37–38, **3**:46–48
digital versatile discs (DVDs) **3**:49, **3**:62
disease **8**:7–8, **8**:23, **8**:26
  and microorganisms **8**:17–21, **8**:36, **8**:56
dishwashers **2**:52, **2**:61
distillation **1**:2–3, **7**:35
DNA *See* deoxyribonucleic acid
domes **2**:16–*17*, **2**:29
drainage **2**:37–38, **2**:49, **4**:10–11, **4**:32
drinks **4**:59–63
drugs **8**:25–33, **8**:57
  immunosupressant **8**:32, **8**:46
  See also herbalism
du Fay, Charles **10**:45–46
Dunlop, John Boyd **6**:21–22
dyes **7**:33, **7**:34, **7**:36
dynamite **2**:49, **7**:7, **9**:24, **9**:25

E

earthquakes **5**:37–38, **5**:39, **5**:40
Eastman, George **3**:27
echolocation **5**:45
Edison, Thomas **3**:17, **3**:30, **3**:33–34, **5**:51, **7**:18, **10**:26, **10**:27, **10**:53

Ehrlich, Paul **8**:27
Einstein, Albert **5**:21, **9**:33, **10**:57
electric currents **10**:44, **10**:46–52
electricity **10**:43–55
  electric light **10**:26–27
  engines **6**:30–31, **6**:33, **6**:46
  measuring **5**:49–51
  static **5**:50, **10**:44, **10**:45
electric motors **10**:51, **10**:52, **10**:53
electroencephalograms **8**:38
electrolysis **7**:28, **7**:37, **7**:41, **10**:41, **10**:48
electromagnetic radiation **3**:18, **3**:41, **5**:26, **5**:27, **5**:50, **6**:35, **10**:48, **10**:49–52
electronic components **10**:47
electroscopes **5**:50
elevators **2**:22–24
enamel **2**:54, **7**:17
endoscopes **8**:43, **8**:52, **8**:53, **8**:60
engines **7**:13, **10**:29–41
  diesel **6**:31, **6**:60, **6**:61, **10**:15, **10**:24, **10**:36–37
  external-combustion **10**:34, **10**:38
  four-stroke **6**:38, **6**:47, **10**:14, **10**:35, **10**:36
  gas turbine **6**:33, **6**:46, **6**:60, **6**:61, **6**:62, **10**:37–40
  internal-combustion **1**:9, **1**:19, **6**:39, **10**:15, **10**:32, **10**:34–36, **10**:38, **10**:40
  ion **1**:63
  jet **1**:27–39, **1**:55, **10**:38, **10**:39
  rocket **1**:33, **1**:34, **1**:43–44, **7**:18, **10**:40
  Stirling **10**:34
  See also steam engines
enzymes, restriction **4**:45–47
Ericsson, John **6**:57, **9**:48, **9**:51
escape velocity **1**:45
ether **8**:15, **8**:16–17
Evans, Oliver **7**:56, **10**:15
explosives **9**:24–25
  See also gunpowder

F
factory farming **4**:36–37
Fahrenheit, Daniel G. **5**:30, **5**:31
Faraday, Michael **8**:16, **10**:49, **10**:51
fax machines **3**:10, **3**:21
feedback systems **7**:58, **7**:59
felt **2**:6, **7**:50
Ferguson tractors **4**:30
fermentation **4**:40–41, **4**:54, **4**:59, **4**:60, **4**:61, **8**:18
Fermi, Enrico **9**:33, **10**:58, **10**:60
fertility treatment **8**:59–60
fertilizers **4**:14, **4**:15, **4**:35, **7**:37
fiberglass **7**:17–18, **8**:52
fiber-optics **3**:13, **3**:20, **3**:23, **3**:48, **3**:62, **8**:43, **8**:52, **8**:53, **8**:60
fibers, synthetic **7**:40–42, **7**:51–53
firearms *See* guns
fishing **4**:20, **4**:21–24
  See also aquaculture
flamethrowers **9**:29
Fleming, Alexander **8**:27
flight **1**:5–39, **10**:13
floods **2**:33, **2**:34–35, **2**:38, **4**:8
flying ball governor **7**:58, **7**:59
flying shuttle **7**:48
food
  and biotechnology **4**:39–41, **4**:48–50
  preservation **4**:54–56
  processing **4**:53–63
  See also agriculture; aquaculture; fishing; hunting

Ford, Henry **4**:30, **6**:43, **7**:57, **7**:58, **7**:61
forging **7**:28
Forlanini, Enrico **6**:62
Formica **7**:11
fortifications **9**:8
Fourdrinier machines **3**:11
four-wheel drive **4**:30, **6**:47
Fox Talbot, Henry **3**:26, **3**:27
Franklin, Benjamin **10**:25, **10**:45
freezing/freezers **2**:55, **2**:56, **4**:56–57
Freud, Sigmund **8**:23
fuel cells **8**:50, **10**:40, **10**:41
fuels **10**:19–25, **10**:58
Fulton, Robert **6**:56, **9**:48, **10**:33
funicular **6**:31
furniture **2**:53

G
Galen **8**:9, **8**:10, **8**:12
Galileo Galilei **5**:19, **5**:25, **5**:30, **8**:37
Galileo space probe **1**:62
galvanization **7**:30
galvanometers **5**:50–51, **8**:38
Garand, John **9**:23
gas (natural) **10**:23, **10**:24–25
  cooking with **2**:53–54
  laws **10**:30
gasoline **10**:22–23
  tractors driven by **4**:28
gas turbines *See* engines, gas turbine
Gatling, Richard **9**:25–26
gears **10**:8–10
  automobile **6**:47, **10**:9
  bicycle **6**:21, **10**:7, **10**:16
Geiger counters **5**:34–35
Gemini missions **1**:47
genes **4**:43, **8**:61, **8**:62
gene therapy **8**:33, **8**:62–63
genetic engineering **4**:25, **4**:34, **4**:42, **4**:45–51, **8**:31, **8**:63
genetics **4**:41–45
genetic screening **8**:60–62
geodesic domes **2**:28–29
Giffard, Henri **1**:8
glass **7**:14–17, **7**:19, **7**:33
gliders **1**:17–18, **1**:22
Global Positioning System **1**:45, **5**:60, **6**:49, **9**:43
global warming **10**:20, **10**:25, **10**:57
Goddard, Robert Hutchings **1**:29, **1**:42, **1**:43
gold **5**:6, **5**:7, **5**:10, **7**:21, **7**:22, **9**:60
Goodyear, Charles **7**:38
grain **4**:11–12, **4**:27–28, **5**:6, **5**:10
Gramme, Zénobe T. **10**:51, **10**:52, **10**:53
gramophones **3**:33–34, **3**:35
graphophones **3**:33
gravity **1**:41
  slingshot **1**:59
  zero **1**:52
Gray, Elisha **3**:16
Greathead, Henry **2**:50–51
greenhouse gases **7**:40, **10**:20
Green Revolution **4**:49
grenades **4**:24, **9**:15, **9**:27
Guericke, Otto von **10**:45, **10**:46
gunpowder **1**:41, **2**:49, **7**:7, **7**:35–36, **9**:5, **9**:14–15, **9**:24, **9**:45, **10**:21
guns **7**:6, **9**:15, **9**:16–19, **9**:24, **9**:51
  hunting with **4**:20, **4**:21
  machine gun **9**:25–26, **9**:30, **10**:15
  See also rifles
Gutenberg, Johannes **3**:5, **3**:8, **3**:11
gynecology **8**:55
gyrocompasses **5**:60
gyroscopes **5**:59, **5**:60

H
Haber–Bosch process **7**:37, **7**:39
Hall, Charles **7**:28
Hargrave, Lawrence **1**:15
harnesses **4**:7, **6**:12–13
Harrison, John **5**:59, **10**:10
harrows **4**:7–8, **4**:30
harvesting **4**:11, **4**:13, **4**:27–29, **4**:53
Harvey, William **8**:10, **8**:45
heart **8**:10, **8**:11
  artificial **8**:51–52
  surgery and transplants **8**:48–51
heart-lung machines **8**:48–49
heat **5**:32
heating **2**:17, **10**:25–26, **10**:41
heat-resistance **1**:53, **7**:18
heavier-than-air craft, development **1**:15–25
helicopters **1**:38–39, **10**:13
Henry, Joseph **10**:51
herbalism **8**:6–7, **8**:25–27, **8**:35, **8**:58
Hero of Alexandria **10**:6, **10**:14, **10**:17, **10**:29–30
Héroult, Paul **7**:28
Herschel, John **3**:25
Hertz, Heinrich **3**:18, **10**:63
Hindenburg (airship) **1**:11–12
Hippocrates **8**:7–8
HIV **8**:28, **8**:58
Hollerith, Herman **3**:54–55
holography **5**:28–30
homes
  green **2**:31
  mechanization in **2**:53, **2**:60–63
  See also buildings
Hooke, Robert **5**:23, **5**:24–25, **5**:40, **5**:41
Hopper, Grace **3**:56
horses **4**:7, **4**:27, **4**:39, **6**:7–15, **6**:30
hospitals, early **8**:6
hovercraft **6**:60, **6**:61, **6**:62–63
Human Genome Project **8**:61–62
humidity measurements **5**:42, **5**:43, **5**:47
humors **8**:35, **8**:36
hunting tools **4**:19–21, **5**:5
Huygens, Christiaan **5**:19, **10**:34
hydraulic systems
  brake **6**:44, **6**:45
  in cranes **2**:15
  and tractors **4**:30
hydroelectricity **2**:37, **10**:54, **10**:61–63
hydrofoils **6**:60–62
hydrogen, powering cars **6**:46
hydroponics **4**:32
hygrometers **5**:42, **5**:43
hypocausts **10**:25–26

I
iconoscopes **3**:43, **3**:44
igloos **2**:8, **2**:10
immune response **8**:20
implants, surgical **8**:51
incubators **8**:56, **8**:57–58
Industrial Revolution **2**:19, **2**:46, **4**:53, **7**:7, **7**:55, **7**:56, **10**:13–15, **10**:16, **10**:21, **10**:26
industry **7**:55, **7**:56–57, **7**:61, **10**:15
insecticides **4**:12, **4**:32–33
integrated circuits/silicon chips **3**:19, **3**:20, **3**:29, **3**:53, **3**:59, **7**:60, **10**:44, **10**:47
interchangeable parts **2**:20, **7**:56–57
Internet **2**:31, **3**:20, **3**:62–63
iron **7**:22–24, **7**:29, **7**:30, **10**:21
  bridges **2**:43, **2**:45–46, **7**:24
  building with **2**:19–22, **7**:25, **7**:26

in ships **6**:57–58, **7**:24, **9**:49–51
  See also steel
irrigation **2**:33, **2**:48, **4**:8–10, **4**:30–32, **10**:6
IVF **8**:59, **8**:60

J
Jacquard loom **3**:52–53, **7**:48–49, **7**:60
jeans **7**:50
Jenner, Edward **8**:19
joints, artificial **8**:51, **8**:52
Joule, James **10**:49
jukeboxes **3**:33
Jupiter, space probes to **1**:60, **1**:62

K
Kalashnikov, Mikhail **9**:23
Kelvin, Baron **5**:31
kerosene **10**:21, **10**:22, **10**:26
Kevlar **1**:62, **7**:18–19, **7**:53
Khorana, Gobind **4**:44–45
kinetoscopes **3**:30
kitchen equipment **2**:53–56, **2**:57, **2**:58, **2**:61
kites **1**:15
knitting **7**:50–51
Koch, Robert **8**:17, **8**:18, **8**:23, **8**:36
Koenig, Friedrich **3**:11
Krupp, Alfred **9**:24

L
lances **9**:14
Land, Edwin **3**:28
Landsteiner, Karl **8**:45
Laplace, Simon **5**:32
lasers **2**:51, **3**:13, **3**:38, **5**:21, **5**:28–30, **5**:40, **9**:63
  laserdiscs **3**:49
  laser printers **3**:10, **3**:11
  for surgery **8**:52, **8**:53
lathes **7**:10–11, **7**:29
latitude **5**:57–58, **5**:61, **5**:62
laundry powders, biological **4**:42
Lavoisier, Antoine **5**:32, **5**:33
Lawes, John Bennett **7**:37
lead **7**:27, **10**:23
Leblanc process **7**:36–37
Leeuwenhoek, Anton van **5**:24, **8**:35
Leibniz, Gottfried **3**:52
Lenoir, Etienne **6**:39
Leonardo da Vinci **1**:15, **1**:16, **5**:42, **6**:18, **9**:30, **10**:13
Le Prince, Louis **3**:29
leprosy **8**:8, **8**:18
levers **7**:56, **10**:5, **10**:6, **10**:7
Lewis gun **9**:26
light **5**:28
  and spectroscopy **5**:27–28
  speed of **5**:10, **5**:21
  See also lasers
light bulbs **7**:28, **10**:26, **10**:27
lighthouses **6**:62
lighting **10**:19, **10**:24, **10**:26–27, **10**:53, **10**:63
Lilienthal, Otto **1**:17
Lindbergh, Charles **1**:21
linen **7**:45
Lippershey, Hans **5**:25
liquid-crystal displays **3**:28, **3**:46
Lister, Joseph **8**:22
lithography **3**:8, **3**:9
locks and keys **9**:57–63
logic gates **3**:57, **10**:47
longitude **5**:58–61, **5**:62
looms **3**:52–53, **7**:47–49, **7**:50–51, **7**:60
lost-wax process **7**:30

loudspeakers **3**:19, **3**:35, **3**:39, **10**:50
Lovelace, Augusta Ada, Countess of **3**:56
Lumiere brothers **3**:30
lunar module **1**:47, **1**:48, **1**:49
lunar roving vehicles **1**:50

M
MacAdam, John **6**:15
McCormick, Cyrus Hall **4**:27–28
maces (clubs) **9**:5–6
machines **7**:56, **10**:5–17
   agricultural **4**:5–8, **4**:11–12, **4**:27–31, **4**:37
   for food production **4**:53
   tunneling **2**:49–50
   See also automation; industry; mechanization
Macintosh, Charles **7**:38, **7**:53
maglev trains **6**:34, **6**:35, **7**:14, **10**:48
magneto optical discs (MODs) **3**:38
magnetophones **3**:35
malaria **8**:18, **8**:26, **10**:20
Malske, Johann Georg **6**:30
maps and charts **5**:56, **5**:61–63
Marconi, Guglielmo **3**:18
Marey, Jules **3**:29–30
margarine **4**:59
Mariner space probes **1**:58–60
Mars **1**:59
   probes to **1**:57, **1**:59, **1**:60, **1**:61, **1**:62, **5**:45, **7**:63
mass spectrometry **5**:34
Matthaei, Heinrich **4**:44–45
Maxim gun **9**:25, **9**:26
Maxwell, James Clerk **3**:18, **10**:51
measurements **5**:5–11, **5**:45–53
   and the globe **5**:55–63
   See also scientific instruments; time
mechanization **7**:55, **7**:56–63
   in the home **2**:52, **2**:60–63
   textile machinery **7**:46–51
   See also machines
medical-imaging equipment **7**:14, **8**:37–43
medicine
   development **8**:5–23
   diagnosis and testing **8**:35–43
   See also drugs; surgery
Mendel, Gregor **4**:41–43
mental illness, drugs **8**:33
Mercator projection **5**:62
Mercury, space probes to **1**:59
metals/metalworking **7**:21–31, **9**:5, **9**:59, **10**:21
micrometer gauges **5**:11
microphones **3**:17, **3**:37, **10**:50
microprocessors **3**:59, **3**:61, **6**:49, **10**:47
microscopes **4**:41, **5**:23–25, **8**:9, **8**:35, **8**:36, **8**:42
   electron **4**:42, **5**:35, **5**:36
microwave ovens **2**:54, **5**:26
milking machines **4**:37
mines/mining **6**:25, **6**:26, **7**:6–9, **7**:39, **7**:61, **9**:31, **10**:30
Minié, Etienne **9**:23
missiles **5**:59, **9**:6, **9**:11–13, **9**:37–42, **9**:43, **9**:63
   arrows **9**:5, **9**:10
   heat-seeking **1**:36, **1**:39
monorails **6**:34
Montgolfier brothers **1**:6–7

Moon **1**:55
   Apollo program **1**:47–49, **3**:46
   phases **5**:14, **5**:16, **5**:17
   probes **1**:58
mordants **7**:34
Morse, Samuel F.B. **3**:15
Morse code **3**:13, **3**:15, **3**:18
mortars **9**:27
motion, laws of **1**:28, **1**:44
motorcycles **6**:38–41
Moulton, Alex **6**:22–23
MRI scans **8**:40–42
multihulls **6**:60, **6**:63
muskets **4**:21, **7**:56, **9**:16, **9**:18–19
Muybridge, Eadweard **3**:27, **5**:27

N
nails **7**:6
napalm **9**:29
NASA **1**:45, **1**:46, **1**:47, **1**:58, **1**:62, **7**:63
   See also Apollo Moon program
natural gas **2**:53, **10**:23
natural selection **4**:40
navigation **5**:6, **5**:55–61
   See also maps and charts
nerve gas **4**:33, **9**:43
networks, computer **3**:62
Newcomen, Thomas **7**:8, **10**:30–32
Newton, Isaac **1**:28, **1**:41
Niépce, Joseph **3**:25
Nightingale, Florence **8**:22, **8**:23
Nipkow, Paul **3**:41
Nirenberg, Marshall **4**:44–45
nitrates **4**:15, **4**:49, **4**:58, **7**:39
nitrites **4**:15, **4**:58
nitrocellulose **7**:39, **7**:40, **7**:51, **7**:52
nitrogen, fixation **4**:15, **4**:49
nitrogen cycle **4**:15
nitrous oxide **8**:15, **8**:16
Nobel, Alfred **7**:7, **9**:24, **9**:25
nuclear power **10**:57–60
   for pacemakers **8**:50
   craft driven by **6**:60, **10**:59, **9**:54, **9**:55
   See also power plants
nuclear reactors **5**:45, **10**:58–59
number systems **5**:7–9
nylon **7**:42, **7**:52, **7**:53

O
obstetrics **3**:18, **5**:50, **8**:55, **10**:49
Ohain, Hans von **1**:30, **10**:39
Ohm, Georg Simon **10**:48
oil **10**:21–22, **10**:23, **10**:24
   spills **6**:59, **10**:57
optical instruments See endoscopes; microscopes; telescopes
orbits **1**:46
organophosphates **4**:33, **4**:34
Ørsted, Hans Christian **3**:15
Otis, Elisha G. **2**:22, **2**:23
Otto, Nikolaus A. **10**:36
Outram, John **6**:30
oxyacetylene torches **7**:31

P
pacemakers **8**:50, **8**:51, **8**:52
paint **7**:34
pants **7**:50, **7**:51
paper **3**:7, **3**:8, **3**:11, **7**:33, **10**:11
Paré, Ambroise **8**:12
Parsons, Charles **6**:58, **10**:37–38

particle accelerators **5**:35–37
Pascal, Blaise **3**:51, **3**:52
pasteurization **4**:61, **8**:18
Pasteur, Louis **4**:41, **4**:61, **8**:17–21, **8**:23, **8**:36, **8**:56
pastoralism **4**:16–17
pendulum clocks **5**:19–20, **5**:25
penicillin **8**:27, **8**:28–30
pest control **4**:34
   pesticides **4**:32–34
petroleum **10**:21–22
PET scans **8**:40, **8**:41, **8**:42
Pfleumer, Fritz **3**:35
Phaistos disk **3**:5
phenakistoscopes **3**:29
phonographs **3**:33, **3**:34, **3**:35
photocopying machines **3**:10
photoelectric effect **10**:63
photography **3**:25–29, **5**:27
piezoelectricity **5**:20–21, **5**:45–46, **10**:50
pikes (weapon) **9**:14
Pioneer space probes **1**:60
plague **4**:17, **8**:8, **8**:18
planets, exploration **1**:57–63
plantations **4**:9, **4**:57
plasmids **4**:47
plastics **7**:33, **7**:39–43, **7**:53
Plimsoll line **6**:56
plows **4**:5–7, **4**:28, **4**:29, **7**:9, **10**:5, **10**:6
Pluto, space probes to **1**:63
pollution **4**:35, **10**:23, **10**:37, **10**:57
polyethylene **7**:42, **7**:43
polymers **7**:42, **7**:43, **7**:52, **7**:53
porcelain **2**:59, **7**:12–13
post-and-lintel system **2**:12–14
pottery **7**:12–13, **10**:8
Poulsen, Valdeman **3**:34–35
power plants **10**:26, **10**:53, **10**:54
   hydroelectric **10**:54, **10**:61, **10**:62
   nuclear **5**:47, **10**:57, **10**:58–60
   solar **10**:63
pregnancy **8**:55–58
   and genetic screening **8**:60–61
   testing **5**:53
   and ultrasound **8**:39
   See also contraception; fertility treatment
printing **3**:5, **3**:7, **3**:8–11
probes, space See space probes
propellers
   on aircraft **1**:18–25, **1**:29
   hovercraft **6**:61
psychiatry **8**:23
psychoanalysis **8**:23
Ptolemy **5**:61–62
puddling process **7**:24
pulleys **2**:15, **2**:23, **4**:23, **7**:56, **7**:57, **10**:5, **10**:6, **10**:7, **10**:17
Pullman railroad cars **2**:59, **6**:28
pyramids **2**:8–9, **2**:10, **2**:53, **5**:14, **5**:55, **7**:5
Pyrex **7**:16, **7**:17
pyrometers **5**:42–43

Q
quartz crystals **5**:20–21, **5**:45–46, **10**:50
quinine **7**:34, **8**:26

R
racing cars **6**:37, **6**:49
radar **1**:35, **1**:49, **1**:62, **5**:43, **5**:60, **5**:61, **9**:37, **9**:38

radio **3**:17–22, **3**:23, **3**:41, **5**:26, **9**:37, **10**:47
radioactivity **5**:26–27, **5**:34–35, **9**:33, **10**:57, **10**:58
radiocarbon dating **5**:49, **5**:50
radiosondes **5**:43, **5**:47
railroads **6**:14, **6**:25–35, **6**:37, **7**:27
   tracks **7**:24
   tunnels **2**:49
rain gauges **5**:40
rally cars **6**:47
ramps **10**:6, **10**:7
rayon **7**:42, **7**:51–52
reaping **4**:11, **4**:27–28
record players and records **3**:35, **3**:37
recycling **7**:11, **7**:28, **7**:40, **7**:42
refrigeration **2**:56, **2**:57, **4**:55–56
reins **6**:9, **6**:10
Reis, Philipp **3**:15–16
relativity, theory of **5**:21
rifles **4**:21, **9**:21–23, **9**:30
roads **6**:14, **6**:15, **6**:48
robots **7**:55, **7**:61–63
rocket engines See engines
rockets **9**:14–15, **9**:31
   Congreve (missiles) **9**:49
   space **1**:41–44, **1**:47, **1**:51, **1**:55, **5**:47, **10**:19
Roebling, John **2**:46
Röntgen, Wilhelm **5**:27
rubber **4**:9, **7**:38
Rutherford, Ernest **9**:33

S
saddles **6**:7, **6**:10, **6**:11
Sagan, Carl **1**:60
sails **4**:22, **6**:52, **6**:53–55
sambucas **9**:9
Santos-Dumont, Alberto **1**:8
satellites (artificial) **1**:42, **1**:45–46, **1**:57–58, **1**:62, **5**:33, **5**:17, **5**:18, **5**:49, **5**:63
   communications **1**:45, **2**:31, **3**:20, **3**:21
   Global Positioning System **1**:45, **5**:60
   military **9**:42–43
   TV broadcasting **3**:48–49
   weather **5**:43
Saturn, space probes to **1**:60
Saturn V **1**:47, **1**:48, **1**:49, **1**:51
saws **2**:47, **7**:6
scaffolding **2**:11
scales **5**:10
scanning, medical **8**:38–42
Schweppe, Jacob **4**:63
scientific instruments **5**:23–43, **10**:9
screw propellers **6**:56–57, **9**:48
screws **7**:6, **10**:7
   auger/Archimedes's **2**:34, **4**:10, **10**:6, **10**:17
sea defenses **2**:38–39
seaplanes **1**:23, **1**:24
seasons **5**:14, **5**:15
seeds, sowing **4**:8, **4**:27
seismographs **5**:37–38, **5**:39
semaphore **3**:14
sewage treatment **2**:37–38
sewing machines **10**:16
Shannon, Claude **3**:56, **3**:57
sheep **4**:14, **4**:16
ships and boats **6**:51–63
   See also warships
shoes **6**:17, **7**:42

shrapnel shells 9:16
siege warfare 9:8–10, 9:15
Siemens-Martin furnace 7:25
Siemens, Werner von 6:30
signals, train 6:29, 6:32, 6:33
Sikorsky, Igor 1:38
silicon chips
 See also integrated circuits
silk 7:45, 7:50
skiing 6:17–18
Skylab project 1:50, 1:51
skyscrapers 2:22–25, 2:30,
 7:26, 7:27
smallpox 8:19
smart materials 7:19
Smeaton, John 6:62
smelting 7:21–23, 10:21
Smith, Francis P. 6:56–57, 9:48
Smith, Oberlin 3:34, 3:35
soap 7:34, 7:36, 7:38
sodium hydroxide 7:37
Sojourner Rover 5:45, 7:63
solar panels 1:53, 1:58, 1:59, 2:31
solar power 10:25, 10:26, 10:63
 for vehicles 6:46, 10:40
Solar System 1:57–63
solar wind 1:59
soldering 7:31
sonar 4:23–24, 5:46, 8:38
sound 10:50
 recording 3:33–39
sound barrier 1:30–33, 1:34,
 1:39, 1:42
space age 1:41–63
spaceplanes 1:55
space probes 1:55, 1:58–63,
 5:45, 7:63
space shuttles 1:40, 1:50, 1:52,
 1:53–55, 2:55, 7:18, 10:41
space stations 1:42, 1:49–53,
 1:55
spacesuits 1:47
spectrometers 5:28, 5:33
spectroscopy 5:27–28
spinning 7:46–47
stars, navigating by 5:56, 5:57
Star Wars 9:41
Stealth aircraft 1:34, 1:35–37
steam engines 4:27, 4:29,
 6:25–27, 7:8, 9:49, 10:14,
 10:15, 10:21, 10:29–34
 feedback systems 7:58, 7:59
steam power
 cranes 2:15
 drills 7:7
 elevators 2:23
 hammers 7:29
 heating 10:26
 locomotives 6:25–27, 10:33
 looms 7:48, 7:50
 pumps 2:34, 2:56, 7:8
 road vehicles 6:37–38
 ships/boats 4:22, 6:55–57, 6:58,
 6:60, 9:48–49, 10:33, 10:38
 See also steam engines
steam turbines 6:58, 10:37–38
steel 6:32, 7:22, 7:24–26, 7:29–30
 bridges 2:43, 2:45–46
 building with 2:25, 2:26, 7:27
 in ships 6:57, 6:58, 9:49
Stephenson, George 6:25–27
Stephenson, Robert 2:43

stethoscopes 8:37–38,
 8:56–57, 9:59
Stevenson screens 5:42
stirrups 6:11
stone, building in 2:8–10, 2:14
Stonehenge 2:12–14, 5:14
streetcars 2:30, 6:30, 10:25
studs 7:52
submarines 9:54, 9:55
subway systems 6:29–30, 6:33
sugar 4:9, 4:57, 4:58
sulfonamides/sulfa drugs
 8:28, 8:57
sulfur 7:35–36
sulfuric acid 7:35, 7:36, 7:37, 7:39
Sun 5:27
 navigating by 5:55, 5:56
sundials 5:18
superconductors 7:14, 10:48
surgery 8:11–13, 8:43, 8:45–53
 and infection 8:22, 8:45
 plastic 8:53
 See also anesthetics
surveying 5:55
suspension 6:15, 6:38, 6:47
sweeteners, artificial 4:57–58
swords 7:22, 9:6, 9:14
Symington, William 6:55–56,
 10:33
syphilis 8:8, 8:18, 8:27

T
tanks 9:21, 9:29, 9:30, 9:31
tape recorders 3:35, 3:36
 digital audio (DAT) 3:38, 3:39
tapes, cassette 3:35, 3:38
tea 4:61–62
telecommunications 3:13–23
 and computers 3:62–63
 See also television
telegraph, electric 3:13–15
telegraphones 3:34–35
telemetry 5:45, 5:47–48, 5:51
telenons 9:10
teleoperators 7:62
telephones 3:15–17, 3:22, 3:23
 mobile phones 3:20–21
telescopes 1:57, 5:23, 5:25
television 3:41–49, 7:63
Telford, Thomas 2:46, 6:15
temperature, measuring 5:30–31,
 5:43, 5:47, 8:9, 8:37
 scales 5:31, 8:36, 8:37
tents 2:5–6, 2:10
textiles, manufacture 7:45–53,
 10:15
TGV trains 6:33
thalidomide 8:33
thermocouples 5:31
thermometers 5:30–31, 5:41–42,
 5:43, 8:9, 8:37
thermostats 10:26
Thomson, Elihu 5:51
threshing 4:11–12, 4:27, 4:53
tides 5:14, 5:16
tiles, heat-resistant 1:53, 7:18
time, measuring 5:13–21
 See also clocks
tin 7:27
tires 6:13, 6:21–22, 6:38,
 6:42, 7:38
tissue culture 4:41

Titanic 6:59
titanium 7:28
titration 5:32
toilets 2:37, 2:55, 2:59–60, 8:7
tomatoes 4:29, 4:32, 4:50
tools 2:8, 7:5–6, 7:9–11,
 7:22, 7:33, 10:5–6
 surgical 8:12, 8:13
torpedoes 5:59, 9:51
Torricelli, Evangelista 5:41
tractors 4:28–30
trains 6:25–35
transducers 5:48, 10:50, 8:39
transfusions, blood 8:45–46
transistors 3:18, 3:45–46, 10:47
 in computers 3:58, 3:59
transplants, organ 4:51, 8:47–50
treadmills 10:10–11, 10:17
trepanning 8:5
Trevithick, Richard 6:25, 6:26,
 6:37, 7:7, 10:33, 10:34
Tsiolkovsky, Konstantin 1:42–43
Tull, Jethro 4:8, 4:14
tungsten 7:28
tunnels 2:47–51
turbines 1:28, 10:38
 hydroelectric 10:61, 10:62
 steam 6:58, 10:37–38
 water 2:61
 wind 10:60
 See also engines, gas turbine
Turing, Alan 3:56, 9:34
typewriters 3:7, 10:16

U
ultrasound 5:45, 5:46,
 8:38, 8:39
ultraviolet light 5:26, 10:27

V
vaccination 8:19, 8:20
vacuum cleaners 2:62–63
vacuum tubes 3:18,
 3:57–58, 10:47
van Allen, James 1:57
Van de Graaf generator 10:44
vaults 2:14–16
velcro 7:52
Venus, probes to 1:58–60, 1:62
Vernier scale 5:11
Vesalius, Andreas 8:10
veterinary science 4:17, 4:35
video recording 3:37, 3:49
Viking space probes 1:57, 1:59,
 1:60, 1:61
Volta, Alessandro 7:28, 10:46,
 10:48
voltage/volts 5:49, 10:44, 10:52
Von Neumann, John 3:58
Voyager space probes 1:60
V/STOL planes 1:34, 1:35
vulcanization 7:38

W
Wallace, Alfred Russel 4:40
warfare
 before 1860 9:5–19
 from 1860–1945 9:21–31
 modern 9:33–43
 sea 9:45–51
warships 7:26, 9:45–52, 10:38,
 10:39, 10:40

washing, clothes 2:60–61
water
 control/use 2:33–39, 10:62
 raising 2:34, 4:10, 10:6, 10:30
 supply 2:33–34, 2:48
 See also drainage; irrigation
water mills 7:47, 7:55, 7:56,
 10:9, 10:11
waterproofing 7:53
water wheels 4:10, 4:27, 10:11
Watson, James 4:44
Watt, James 5:9, 7:58, 10:30,
 10:32–34
weapons See warfare
weather measurements 5:38–43
weather stations 5:42, 5:43
weaving 7:45, 7:47–49, 7:50
Wedgwood, Josiah 7:13
weightlessness 1:52
weights 5:5, 5:6–7
welding 7:31, 7:62, 10:25
whaling 4:20, 4:24
Wheatstone, Charles 3:14–15
wheels 6:6–7, 6:8, 6:32,
 7:56, 10:6–10
 See also gears
whiskey 4:60–61
Whitney, Eli 7:46, 7:56–57
Whittle, Frank 1:27, 1:29–30,
 10:39
widgets 4:60
Wilkinson, John 6:55
windmills 7:55, 10:9, 10:11–12
wind power 10:40, 10:60–61,
 10:63
wind speed 5:38
wine 4:60, 4:61
winnowing 4:11, 4:12
wood 10:19–20
woodworking 7:9–12
wool 7:45
World War I
 airships 1:10–11
 communications 9:36–37
 deaths 9:30
 tanks 9:21, 9:30
 warplanes 1:20
 weapons 9:24, 9:26, 9:27,
 9:28, 9:29
World War II
 airplanes 1:21–25, 1:30, 1:33
 atomic bombs 9:33, 9:34
 barrage balloons 1:13
 communications 9:36–37
 gas masks 9:28
 rockets 1:42–44
 tanks 9:29, 9:30
 weapons 9:23, 9:25, 9:29, 9:31
Wright, Frank Lloyd 2:28
Wright brothers 1:17–19, 1:20,
 10:39
writing 3:5–7

X, Y, and Z
xerography 3:10
X-rays 5:27, 8:37, 8:38, 8:39
Yale, Linus 9:60
zeppelins 1:9–11
zippers 7:52
zoetropes 3:29
Zuse, Konrad 3:56
Zworykin, Vladimir 3:43, 3:44